# THERMO-HYDRAULICS OF NUCLEAR REACTORS

This book provides a concise and up-to-date summary of the essential thermo-hydraulic analyses and design principles of nuclear reactors for electricity generation. Beginning with the basic nuclear physics, it leads through technical and quantitative analyses to descriptions of both the normal operation of the various modern nuclear reactor designs and the analyses of the possible departures from normal operation. It then describes both the postulated accident scenarios and summaries of the causes for the three major nuclear power generation accidents, Three Mile Island, Chernobyl, and Fukushima, as well as the major improvements to reactor safety that grew out of those analyses and accidents.

Professor Christopher Earls Brennen was a member of the Mechanical Engineering Faculty at Caltech for over 40 years and retired as the Richard L. and Dorothy M. Hayman Professor of Mechanical Engineering in 2005. As a teacher he was the recipient of a number of teaching awards including the prestigious Richard Feynman Prize.

# Thermo-Hydraulics of Nuclear Reactors

Christopher Earls Brennen

*California Institute of Technology*

# CAMBRIDGE
## UNIVERSITY PRESS

32 Avenue of the Americas, New York, NY 10013-2473, USA

Cambridge University Press is part of the University of Cambridge.

It furthers the University's mission by disseminating knowledge in the pursuit of education, learning, and research at the highest international levels of excellence.

www.cambridge.org
Information on this title: www.cambridge.org/9781107139602

First published 2016

Printed in the United States of America

*A catalog record for this publication is available from the British Library.*

*Library of Congress Cataloging in Publication Data*
Names: Brennen, Christopher E. (Christopher Earls), 1941– author.
Title: Thermo-hydraulics of nuclear reactors / Christopher Earls Brennen,
    California Institute of Technology Pasadena, California.
Description: New York, NY : Cambridge University Press, [2016] | !!2014 |
    Includes bibliographical references and index.
Identifiers: LCCN 2015039287 | ISBN 9781107139602 (hardback ; alk. paper) |
    ISBN 1107139600 (hardback ; alk. paper)
Subjects: LCSH: Nuclear reactors–Fluid dynamics. | Nuclear energy.
Classification: LCC TK9202 .B66 2016 | DDC 621.48/3–dc23
LC record available at http://lccn.loc.gov/2015039287

ISBN 978-1-107-13960-2 Hardback

# Contents

*Preface*                                                                    *page* ix

*Mathematical Nomenclature*                                                        xi

1  **Introduction** . . . . . . . . . . . . . . . . . . . . . . . . . . . . . . . . . 1
   1.1  Background and Context                                          1
   1.2  This Book                                                       2

2  **Basic Nuclear Power Generation** . . . . . . . . . . . . . . . . . . . 5
   2.1  Nuclear Power                                                   5
   2.2  Nuclear Fuel Cycle                                              5
      2.2.1  Thorium Fuel Cycle                                   7
      2.2.2  Fuel Changes in the Reactor                          8
      2.2.3  The Postreactor Stages                               8
   2.3  Nuclear Physics                                                 9
      2.3.1  Basic Nuclear Fission                                9
      2.3.2  Neutron Energy Spectrum                             10
      2.3.3  Cross Sections and Mean Free Paths                  11
      2.3.4  Delayed Neutrons and Emissions                      13
   2.4  Radioactivity and Radioactive Decay                            13
      2.4.1  Half-Life                                           13
      2.4.2  Decay in a Nuclear Reactor                          14
   2.5  Radiation                                                      15
   2.6  Containment Systems                                            16
      2.6.1  Radioactive Release                                 16
      2.6.2  Reactor Shielding                                   17
   2.7  Natural Uranium Reactors                                       18
   2.8  Thermal Reactors                                               18
      2.8.1  Moderator                                           18
      2.8.2  Neutron History in a Thermal Reactor                20

v

2.9   Fast Reactors                                                      21
2.10  Criticality                                                        21
2.11  Fuel Cycle Variations                                              22

3  **Core Neutronics** . . . . . . . . . . . . . . . . . . . . . . . . . . . 25
   3.1   Introduction                                                     25
   3.2   Neutron Density and Neutron Flux                                 25
   3.3   Discretizing the Energy or Speed Range                           26
   3.4   Averaging over Material Components                               27
   3.5   Neutron Transport Theory                                         28
   3.6   Diffusion Theory                                                 30
         3.6.1  Introduction                                              30
         3.6.2  One-Speed and Two-Speed Approximations                    32
         3.6.3  Steady State One-Speed Diffusion Theory                   33
         3.6.4  Two-Speed Diffusion Theory                                34
         3.6.5  Nonisotropic Neutron Flux Treatments                      36
         3.6.6  Multigroup Diffusion Theories and Calculations            36
         3.6.7  Lattice Cell Calculations                                 37
   3.7   Simple Solutions to the Diffusion Equation                       37
         3.7.1  Spherical and Cylindrical Reactors                        37
         3.7.2  Effect of a Reflector on a Spherical Reactor              40
         3.7.3  Effect of a Reflector on a Cylindrical Reactor            42
         3.7.4  Effect of Control Rod Insertion                           43
   3.8   Steady State Lattice Calculations                               45
         3.8.1  Introduction                                              45
         3.8.2  Fuel Rod Lattice Cell                                     47
         3.8.3  Control Rod Lattice Cell                                  49
         3.8.4  Other Lattice Scales                                      50
   3.9   Unsteady or Quasi-Steady Neutronics                             51
         3.9.1  Unsteady One-Speed Diffusion Theory                      51
         3.9.2  Point Kinetics Model                                      53
   3.10  More Advanced Neutronic Theory                                  53
   3.11  Monte Carlo Calculations                                        54

4  **Some Reactor Designs** . . . . . . . . . . . . . . . . . . . . . . . . 56
   4.1   Introduction                                                     56
   4.2   Current Nuclear Reactors                                         56
   4.3   Light Water Reactors (LWRs)                                      57
         4.3.1  Types of LWRs                                             57
         4.3.2  Pressurized Water Reactors (PWRs)                         58
         4.3.3  Boiling Water Reactors (BWRs)                             60
         4.3.4  Fuel and Control Rods for LWRs                            62
         4.3.5  Small Modular Reactors                                    65
         4.3.6  LWR Control                                               66

4.4 Heavy Water Reactors (HWRs)      67
4.5 Graphite-Moderated Reactors      69
4.6 Gas-Cooled Reactors      69
4.7 Fast Neutron Reactors (FNRs)      70
4.8 Liquid Metal Fast Breeder Reactors      70
4.9 Generation IV Reactors      74
     4.9.1 Generation IV Thermal Reactors      75
     4.9.2 Generation IV Fast Reactors      76

5 **Core Heat Transfer** . . . . . . . . . . . . . . . . . . . . . . . . . . . 78
5.1 Heat Production in a Nuclear Reactor      78
     5.1.1 Introduction      78
     5.1.2 Heat Source      78
     5.1.3 Fuel Rod Heat Transfer      79
     5.1.4 Heat Transfer to the Coolant      82
5.2 Core Temperature Distributions      83
5.3 Core Design: An Illustrative LWR Example      84
5.4 Core Design: An LMFBR Example      85
5.5 Boiling Water Reactor      86
     5.5.1 Temperature Distribution      86
     5.5.2 Mass Quality and Void Fraction Distribution      87
5.6 Critical Heat Flux      89

6 **Multiphase Flow** . . . . . . . . . . . . . . . . . . . . . . . . . . . . 90
6.1 Introduction      90
6.2 Multiphase Flow Regimes      90
     6.2.1 Multiphase Flow Notation      90
     6.2.2 Multiphase Flow Patterns      91
     6.2.3 Flow Regime Maps      92
     6.2.4 Flow Pattern Classifications      93
     6.2.5 Limits of Disperse Flow Regimes      95
     6.2.6 Limits on Separated Flow      96
6.3 Pressure Drop      99
     6.3.1 Introduction      99
     6.3.2 Horizontal Disperse Flow      99
     6.3.3 Homogeneous Flow Friction      100
     6.3.4 Frictional Loss in Separated Flow      101
6.4 Vaporization      105
     6.4.1 Classes of Vaporization      105
     6.4.2 Homogeneous Vaporization      105
     6.4.3 Effect of Interfacial Roughness      108
6.5 Heterogeneous Vaporization      108
     6.5.1 Pool Boiling      108
     6.5.2 Pool Boiling on a Horizontal Surface      109

|       | 6.5.3 | Nucleate Boiling | 111 |
|       | 6.5.4 | Pool Boiling Crisis | 113 |
|       | 6.5.5 | Film Boiling | 115 |
|       | 6.5.6 | Boiling on Vertical Surfaces | 116 |
| 6.6   | Multiphase Flow Instabilities | | 118 |
|       | 6.6.1 | Introduction | 118 |
|       | 6.6.2 | Concentration Wave Oscillations | 119 |
|       | 6.6.3 | Ledinegg Instability | 119 |
|       | 6.6.4 | Chugging and Condensation Oscillations | 120 |
| 6.7   | Nuclear Reactor Context | | 124 |

**7  Reactor Multiphase Flows and Accidents** . . . . . . . . . . . . . . . . . **127**

| 7.1   | Multiphase Flows in Nuclear Reactors | | 127 |
|       | 7.1.1 | Multiphase Flow in Normal Operation | 127 |
|       | 7.1.2 | Void Fraction Effect on Reactivity | 128 |
|       | 7.1.3 | Multiphase Flow during Overheating | 128 |
| 7.2   | Multiphase Flows in Nuclear Accidents | | 130 |
| 7.3   | Safety Concerns | | 130 |
| 7.4   | Safety Systems | | 131 |
|       | 7.4.1 | PWR Safety Systems | 132 |
|       | 7.4.2 | BWR Safety Systems | 133 |
| 7.5   | Major Accidents | | 134 |
|       | 7.5.1 | Three Mile Island | 134 |
|       | 7.5.2 | Chernobyl | 137 |
|       | 7.5.3 | Fukushima | 140 |
|       | 7.5.4 | Other Accidents | 141 |
| 7.6   | Hypothetical Accident Analyses | | 142 |
|       | 7.6.1 | Hypothetical Accident Analyses for LWRs | 142 |
|       | 7.6.2 | Loss-of-Coolant Accident: LWRs | 143 |
|       | 7.6.3 | Loss-of-Coolant Accident: LMFBRs | 145 |
|       | 7.6.4 | Vapor Explosions | 146 |
|       | 7.6.5 | Fuel–Coolant Interaction | 147 |
| 7.7   | Hypothetical Accident Analyses for FBRs | | 147 |
|       | 7.7.1 | Hypothetical Core Disassembly Accident | 148 |

*Index*                                                                           151

# Preface

This book presents an overview of the thermo-hydraulics of the nuclear reactors designed to produce power using nuclear fission. The book began many years ago as a series of notes prepared for a graduate student course at the California Institute of Technology. When, following the Three Mile Island accident in 1979, nuclear power became politically unpopular, demand and desire for such a course waned, and I set the book aside in favor of other projects. However, as the various oil crises began to accentuate the need to explore alternative energy sources, the course and the preparation of this book were briefly revived. Then came the terrible Chernobyl accident in 1986, and the course and the book got shelved once more. However, the pendulum swung back again as the problems of carbon emissions and global warming rose in our consciousness and I began again to add to the manuscript. Even when the prospects for nuclear energy took another downturn in the aftermath of the Fukushima accident (in 2011), I decided that I should finish the book whatever the future might be for the nuclear power industry. I happen to believe, despite the accidents – or perhaps because of them – that nuclear power will be an essential component of electricity generation in the years ahead.

The book is an introduction to a graduate-level (or advanced undergraduate-level) course in the thermo-hydraulics of nuclear power generation. Because neutronics and thermo-hydraulics are closely linked, a complete understanding of thermo-hydraulics and the associated safety issues also requires knowledge of the neutronics of nuclear power generation and, in particular, of the interplay between the neutronics and the thermo-hydraulics that determine the design of the reactor core. This material necessarily leads into the critical issues associated with nuclear reactor safety, and this, in turn, would be incomplete without brief descriptions of the three major accidents (Three Mile Island, Chernobyl, and Fukushima) that have influenced the development of nuclear power.

Some sections in Chapter 6 of this book were adapted from two of my other books, *Cavitation and Bubble Dynamics* and *Fundamentals of Multiphase Flow*, and I am grateful to the publisher of those books, Cambridge University Press, for permission to reproduce those sections and their figures in the present text. Other figures and photographs reproduced in this book are acknowledged in their respective captions. I would also like to express my gratitude to the senior colleagues at the California Institute of Technology who introduced me to the topic of nuclear power generation, in particular, Noel Corngold and Milton Plesset. Milton did much to advance the cause of nuclear power generation in the United States, and I am much indebted to him for his guidance. I also appreciate the interactions I had with colleagues at other institutions, including Ivan Catton, the late Ain Sonin, George Maise, and the staff at the Nuclear Regulatory Commission.

This book is dedicated to James MacAteer, from whom I first heard the word *neutron*, and to the Rainey Endowed School in Magherafelt, where the physics Johnny Mac taught me stayed with me throughout my life.

California Institute of Technology, November 2013

# Mathematical Nomenclature

## Roman letters

| | |
|---|---|
| $a$ | Amplitude of wave-like disturbance |
| $A$ | Cross-sectional area |
| $A$ | Atomic weight |
| $b$ | Thickness |
| $B_g^2$ | Geometric buckling |
| $B_m^2$ | Material buckling |
| $c$ | Speed of sound |
| $c_p$ | Specific heat of the coolant |
| $C, C_1, C_2, C_R$ | Constants |
| $C^*, C^{**}$ | Constants |
| $C_f$ | Friction coefficient |
| $C_i$ | Concentration of precursor $i$ |
| $d$ | Diameter |
| $D$ | Neutron diffusion coefficient |
| $D_h$ | Hydraulic diameter of coolant channel |
| $E$ | Neutron kinetic energy |
| $E'$ | Neutron energy prior to scattering |
| $f$ | Frequency |
| $g$ | Acceleration due to gravity |
| $h, h^*$ | Heat transfer coefficients |
| $H$ | Height |
| $H_E$ | Extrapolated height |
| $Hm$ | Haberman-Morton number, normally $g\mu^4/\rho S^3$ |
| $j$ | Total volumetric flux |
| $j_N$ | Volumetric flux of component $N$ |
| $J_j$ | Angle-integrated angular neutron current density vector |
| $J_j^*$ | Angular neutron current density vector |
| $k$ | Multiplication factor |
| $k_\infty$ | Multiplication factor in the absence of leakage |

| | |
|---|---|
| $k$ | Thermal conductivity |
| $\mathcal{K}$ | Frictional constants |
| $l$ | Typical dimension of a reactor |
| $\ell$ | Typical dimension |
| $\ell$ | Mean free path |
| $\ell_a$ | Mean free path for absorption |
| $\ell_f$ | Mean free path for fission |
| $\ell_s$ | Mean free path for scattering |
| $L$ | Neutron diffusion length, $(D/\Sigma_a)^{\frac{1}{2}}$ |
| $\mathcal{L}$ | Latent heat of vaporization |
| $\dot{m}$ | Mass flow rate |
| $m$ | Index denoting a core material |
| $M$ | Number of different core materials denoted by $m = 1$ to $M$ |
| $Ma$ | Square root of the Martinelli parameter |
| $n$ | Integer |
| $n(E)dE$ | Number of neutrons with energies between $E$ and $E + dE$ |
| $N$ | Number of neutrons or nuclei per unit volume |
| $N_f$ | Number of fuel rods |
| $\mathcal{N}$ | Number of atoms per unit volume |
| $N^*$ | Site density, number per unit area |
| $Nu$ | Nusselt number, $hD_h/k_L$ |
| $p$ | Pressure |
| $p^T$ | Total pressure |
| $P$ | Power |
| $\mathcal{P}$ | Perimeter |
| $(1 - P_F)$ | Fraction of fast neutrons that are absorbed in $^{238}U$ |
| $(1 - P_T)$ | Fraction of thermal neutrons that are absorbed in $^{238}U$ |
| $Pr$ | Prandtl number |
| $\dot{q}$ | Heat flux per unit surface area |
| $\mathcal{Q}$ | Rate of heat production per unit length of fuel rod |
| $r$ | Radial coordinate |
| $r, \theta, z$ | Cylindrical coordinates |
| $R$ | Radius of reactor or bubble |
| $R_E$ | Extrapolated radius |
| $R_R$ | Reflector outer radius |
| $R_{RE}$ | Extrapolated reflector radius |
| $R_P$ | Fuel pellet radius |
| $R_O$ | Outer radius |
| $Re$ | Reynolds number |
| $s$ | Coordinate measured in the direction of flow |
| $S(x_i, t, E)$ | Rate of production of neutrons of energy, $E$, per unit volume |
| $\mathcal{S}$ | Surface tension |
| $t$ | Time |

| | |
|---|---|
| $T$ | Temperature |
| $u, U$ | Velocity |
| $\bar{u}$ | Neutron velocity |
| $u_i$ | Fluid velocity vector |
| $u_N$ | Fluid velocity of component $N$ |
| $V$ | Volume |
| $\dot{V}$ | Volume flow rate |
| $x, y, z$ | Cartesian coordinates |
| $x_i$ | Position vector |
| $x_N$ | Mass fraction of component $N$ |
| $\mathcal{X}$ | Mass quality |
| $z$ | Elevation |

## Greek letters

| | |
|---|---|
| $\alpha$ | Volume fraction |
| $\alpha_L$ | Thermal diffusivity of liquid |
| $\alpha_{mf}$ | Ratio of moderator volume to fuel volume |
| $\beta$ | Fractional insertion |
| $\beta$ | Volume quality |
| $\beta$ | Fraction of delayed neutrons |
| $\epsilon$ | Fast fission factor of $^{238}U$ |
| $\delta$ | Boundary layer thickness |
| $\eta$ | Efficiency |
| $\eta$ | Thermal fission factor of $^{238}U$ |
| $\theta$ | Angular coordinate |
| $\kappa$ | Bulk modulus of the liquid |
| $\kappa$ | Wave number |
| $\kappa_L, \kappa_G$ | Shape constants |
| $\lambda$ | Wavelength |
| $\lambda_i$ | Decay constant of precursor $i$ |
| $(1 - \Lambda_F)$ | Fraction of fast neutrons that leak out of the reactor |
| $(1 - \Lambda_T)$ | Fraction of thermal neutrons that leak out of the reactor |
| $\xi$ | Time constant |
| $\xi_1, \xi_2$ | Constants |
| $\mu, \nu$ | Dynamic and kinematic viscosity |
| $\rho$ | Density |
| $\rho$ | Reactivity, $(k - 1)/k$ |
| $\sigma$ | Cross section |
| $\sigma_a, \sigma_f, \sigma_s$ | Cross sections for absorption, fission, and scattering |
| $\Sigma$ | Macroscopic cross section, $\mathcal{N}\sigma$ |
| $\Sigma_{tr}$ | Macroscopic transport cross section, $1/3D$ |

| | |
|---|---|
| $\tau$ | Half-life |
| $\tau_w$ | Wall shear stress |
| $\phi$ | Angle-integrated neutron flux |
| $\phi_L^2, \phi_G^2, \phi_{L0}^2$ | Martinelli pressure gradient ratios |
| $\varphi$ | Angular neutron flux |
| $\omega$ | Radian frequency |
| $\omega_a$ | Acoustic mode radian frequency |
| $\omega_m$ | Manometer radian frequency |
| $\Omega_j$ | Unit direction vector |

## Subscripts

On any variable, $Q$:

| | |
|---|---|
| $Q_o$ | Initial value, upstream value, or reservoir value |
| $Q_1, Q_2$ | Values at inlet and discharge |
| $Q_a$ | Pertaining to absorption |
| $Q_b$ | Bulk value |
| $Q_c$ | Critical values and values at the critical point |
| $Q_d$ | Detachment value |
| $Q_e$ | Effective value or exit value |
| $Q_e$ | Equilibrium value or value on the saturated liquid-vapor line |
| $Q_i$ | Components of vector $Q$ |
| $Q_f$ | Pertaining to fission or a fuel pellet |
| $Q_s$ | Pertaining to scattering |
| $Q_w$ | Value at the wall |
| $Q_A, Q_B$ | Pertaining to general phases or components, $A$ and $B$ |
| $Q_B$ | Pertaining to the bubble |
| $Q_C$ | Pertaining to the continuous phase or component, $C$ |
| $Q_C$ | Critical value |
| $Q_C$ | Pertaining to the coolant or cladding |
| $Q_{CI}$ | Pertaining to the inlet coolant |
| $Q_{CS}$ | Pertaining to the inner cladding surface |
| $Q_D$ | Pertaining to the disperse phase or component, $D$ |
| $Q_E$ | Equilibrium value |
| $Q_F$ | Pertaining to fast neutrons |
| $Q_{FS}$ | Pertaining to the fuel pellet surface |
| $Q_G$ | Pertaining to the gas phase or component |
| $Q_L$ | Pertaining to the liquid phase or component |
| $Q_M$ | Mean or maximum value |
| $Q_N$ | Nominal conditions or pertaining to nuclei |
| $Q_N$ | Pertaining to a general phase or component, $N$ |
| $Q_R$ | Pertaining to the reflector |

| | |
|---|---|
| $Q_S$ | Pertaining to the surface |
| $Q_T$ | Pertaining to thermal neutrons |
| $Q_V$ | Pertaining to the vapor |
| $Q_\infty$ | Pertaining to conditions far away |

## Superscripts and other qualifiers

On any variable, $Q$:

| | |
|---|---|
| $\bar{Q}$ | Mean value of $Q$ |
| $\dot{Q}$ | Time derivative of $Q$ |
| $\delta Q$ | Small change in $Q$ |
| $\Delta Q$ | Difference in $Q$ values |
| $Q^m$ | Pertaining to the material component, $m$ |

# 1

## Introduction

### 1.1 Background and Context

Beginning in the early 1950s, the nuclear power industry in the United States grew to become second only to coal in its electrical generation capacity. By 1990, there were 111 commercial nuclear power plants with a combined capacity of 99,000 MW, representing about 19 percent of the nation's electric power. Nuclear power production in the United States was then $530 \times 10^9$ kWh, much more than in France and Japan combined, although these two countries were among the nations most reliant on nuclear power. France produced 77 percent of its electricity by nuclear power; in West Germany and Japan, the percentages were 33 percent and 26 percent, respectively. However, in the United States, no new nuclear plants were ordered after 1978, and the expansion of the U.S. commercial nuclear power industry ceased shortly thereafter. Other countries saw a similar drastic decline in the growth of nuclear power capacity.

The reasons for this abrupt transition are several. First, the rate of growth of demand for electric power was less than expected. Second, the capital costs associated with new nuclear power plants rose dramatically in the 1970s and 1980s, in part because of more stringent regulatory activity. And third, public opposition to nuclear power also rose substantially in the aftermath of the Three Mile Island accident (see Section 7.5.1) in 1979, a reaction that was further amplified by the Chernobyl accident in 1986 (see Section 7.5.2). These accidents greatly heightened the public fear of nuclear power plants based on three major concerns, two reasonable and one unreasonable. The unreasonable concern was that a nuclear generating plant might explode like a nuclear weapon, an event that can be dismissed on fundamental physical grounds (see, e.g., Nero 1979). However, the other two concerns that continue to have validity are the fear of the release of harmful radioactive material and the concern over the storage of nuclear waste. While Chernobyl rightly increased the concern over radioactive release, the improvements introduced as a result of the lessons learned from the nuclear accidents over the past half-century (see Sections 7.5 and 7.6) have greatly reduced the risk of such events. Specifically, it is now recognized that, in the past, a lack of standardization in the design and operation of nuclear power plants significantly impaired their safety margins and that worldwide cooperation, oversight, and standardization will radically improve safety margins in the future. Great strides have been made in this regard since the end of the Cold War. Similarly, plans for waste storage and/or recycling continue to be developed both nationally and globally. As von Hippel (2006) has pointed out, there is no hurry to recycle nuclear waste because many temporary storage options are possible given how small a volume of waste is produced, and temporary storage is advisable when a number of reprocessing options may be found to be advantageous in the years ahead.

Of course, no power-generating process is devoid of risks and consequences, and, although complex, it is necessary to balance both the long- and short-term effects while seeking an appropriate mix of energy resources. In 2011, 63 percent of the world's electricity generation was produced by coal and gas combustion; 12 percent was from nuclear power (Shift Project Data Portal 2011). This 12 percent is significantly smaller than in the year 2006, when nuclear power amounted to about 20 percent of global generation. It is projected that nuclear power generation will remain relatively constant in the decades ahead, while the overall demand and generation will continue to grow. This growth is in part caused by population increase and in part by economic development, particularly in the developing countries. Efforts to conserve energy in the developed countries have been more than offset by population increases in the less-developed world. Consequently, worldwide energy consumption per capita continues to rise and increased by approximately 20 percent between 1980 and 2010 (Shift Project Data Portal 2011).

However, it is now becoming clear that the increase in the use of combustible fuels, primarily coal and gas, has serious consequences for the earth's atmosphere and climate, because worldwide emissions of $CO_2$ from electricity production will continue to rise in the decade ahead. Moreover, greenhouse gas emissions are primarily caused by the burning of the combustible fuels coal, natural gas, and oil, which far exceeds that from the other power sources. The emissions advantage of nuclear power generation has led a number of environmental groups to begin to advocate for nuclear power (see, e.g., Duffey et al. 2006) as a preferred *green solution* to the energy challenge. Whatever the preferred means of electricity production might be in the future, it seems clear that nuclear power must remain an option. One of the disturbing consequences of the antinuclear public sentiment in the past 30 years is that nuclear engineering has become quite unpopular in universities (at least in the United States), and hence the numbers of nuclear engineering programs and their students dwindled. If nuclear power generation were to become an important national or global objective, there would have to be a radical increase in that component of our engineering educational effort.

## 1.2  This Book

This book, which is intended as an introduction to the thermo-hydraulics of nuclear power generation for graduates or advanced undergraduates, clearly focuses on just one aspect of the design of nuclear reactors for electricity generation, namely, thermo-hydraulics and issues that affect thermo-hydraulics. The term *thermo-hydraulics* refers to all the flow processes involved in the removal of heat generated in the reactor core and the use of that heat to drive generators that produce electricity. Note that although the use of the word *hydraulics* might imply only water flows, in fact the fluids involved range over many coolants and their liquid and vapor phases, including complex multiphase flows. In the present context, the word *thermo-hydraulics* also refers to a whole collection of possible flow processes that might occur due not only to normal reactor operation but also to any operational irregularities or accidents.

Clearly, then, any review or analysis of the thermo-hydraulics must include a description of how the heat is generated within the nuclear reactor core and, consequently, must include description and quantification of the nuclear physics processes that generate the heat. Thus, following a brief introduction of the background and context of nuclear power generation, Chapter 2 provides a review of the fundamental physics of nuclear fission and radioactivity. This leads into Chapter 3, which covers some of the basic features of the neutronics of nuclear reactors. This is followed in Chapter 4 by a description of the structure of the fission reactors presently used or envisaged for nuclear power generation. With that structure in mind, the reader is then equipped to absorb, in Chapter 5, how the heat generated by nuclear fission is transferred to the reactor core coolant and thus transported out of the core to be used to drive the turbines and generators that complete the structure of the power station. Chapter 6 reviews some of the basic multiphase flow phenomena that may be associated with those heat transfer processes during both normal operation of a nuclear power plant and during postulated accidents within that reactor. This leads naturally to a discussion in Chapter 7 of nuclear reactor safety, including descriptions of the three major accidents that dominate the public's impression of the dangers of nuclear power, namely, the accidents at Three Mile Island, Chernobyl, and Fukushima. That discussion naturally includes the important lessons learned from those accidents and other experiences.

There are, of course, many fine textbooks on nuclear power generation and on the engineering of nuclear power systems (see, e.g., Gregg King 1964). Those interested in more detailed treatments of the analytical methods should consult one of the classic texts, such as Glasstone and Sesonske (1981) or Duderstadt and Hamilton (1976). Other texts, such as Winterton (1981) or Collier and Hewitt (1987), have strong focus on thermo-hydraulics. Of course, many additional aspects associated with nuclear power are also important, such as waste disposal (see, e.g., Knief 1980) and the political and economic issues. Other texts are referenced at the conclusion of each chapter. Moreover, today a great deal can be learned from the pages of the Internet, for example, those constructed by the American Nuclear Society or the World Nuclear Association (WNA 2011). Indeed, any single book attempting to review the entire field of electricity generation by nuclear power would be huge; even many of the more narrowly focused books include excessive detail. The present text attempts to narrow thermo-hydraulics down to its essentials without eliminating essential analytical and practical approaches.

## REFERENCES

Collier, J. G., and Hewitt, G. F. (1987). *Introduction to nuclear power.* Hemisphere.

Duderstadt, J. J., and Hamilton, L. J. (1976). *Nuclear reactor analysis.* John Wiley.

Duffey, R. B., Torgerson, D. F., Miller, A. I., and Hopwood, J. (2006). Canadian solutions to global energy and environment challenges: Green atoms. *EIC Climate Change Conference,* 1–7.

Glasstone, S., and Sesonske, A. (1981). *Nuclear reactor engineering.* Van Nostrand Reinhold. [See also Glasstone (1955) and Bell and Glasstone (1970)]

Gregg King, C. D. (1964). *Nuclear power systems.* Macmillan.

Knief, R. A. (1992). *Nuclear engineering: Theory and practice of commercial nuclear power.* Hemisphere.

Nero, A. V. (1979). *A guidebook to nuclear reactors.* University of California Press.

Shift Project Data Portal. (2011). http://www.tsp-data-portal.org/.

von Hippel, F. N. (2006). No hurry to recycle. *ASME Mechanical Engineering,* **128**, 32–35.

Winterton, R. H. S. (1981). *Thermal design of nuclear reactors.* Pergamon Press.

WNA. (2011). *World Nuclear Association Information Library.* http://www.world-nuclear.org/Information-Library/.

# 2

## Basic Nuclear Power Generation

### 2.1 Nuclear Power

Nuclear energy is released when atoms are either split into smaller atoms (a phenomenon known as fission) or combined to form a larger atom (a phenomenon known as fusion). This monograph will focus on the production of power by harnessing atomic fission since that is the principle process currently utilized in man-made reactors.

Most of the energy produced by nuclear fission appears as heat in the nuclear reactor core, and this heat is transported away from the core by conventional methods, namely, by means of a cooling liquid or gas. The rest of the power generation system is almost identical in type to the way in which heat is utilized in any other generating station, whether powered by coal, oil, gas, or sunlight. Often the heat is used to produce steam that is then fed to a steam turbine that drives electric generators. In some plants, hot gas rather than steam is used to drive the turbines. In the case of steam-generating nuclear power plants, the part of the plant that consists of the reactor and the primary or first-stage cooling systems (pumps, heat exchangers, etc.) is known as the *nuclear steam supply system*, and the rest, the conventional use of the steam, is called the *balance of plant*. This monograph does not deal with this conventional power generation technology but focuses on the nuclear reactor, its production of heat, and the primary coolant loop that cools the reactor core.

### 2.2 Nuclear Fuel Cycle

Though it is possible that power might be derived from nuclear fusion at some point in the distant future, all presently feasible methods of nuclear power generation utilize the energy released during nuclear fission, that is to say, the process by which a neutron colliding with an atom causes that atom to split and, as a by-product, produces heat. With atoms known as fissile atoms, additional neutrons are released at the same time, thus allowing a continuing, naturally regenerating process of fission and a source of heat. The only naturally occurring fissile material is the uranium isotope, $^{235}$U, but it only occurs along with a much greater quantity of the common

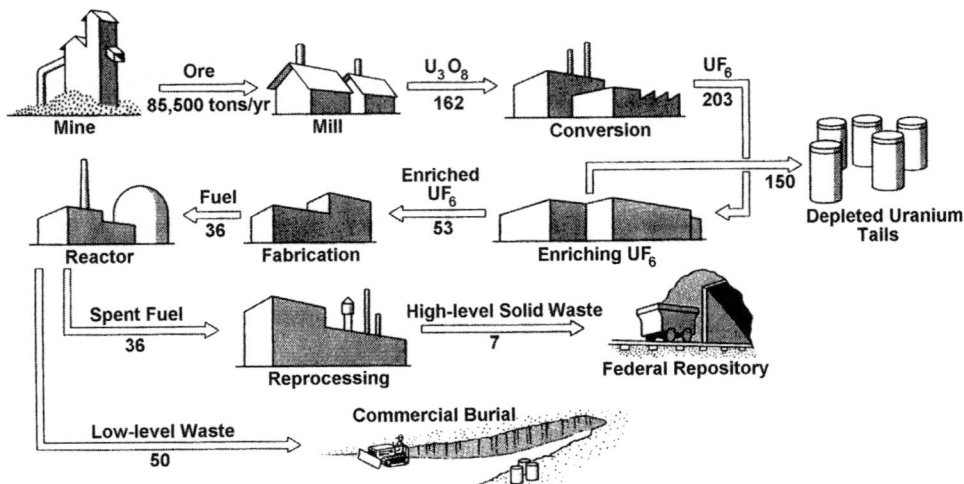

Figure 2.1. Uranium requirements for a typical pressurized water nuclear reactor (see Section 4.3.2). The numbers refer to the number of tons of each material required per year for a 1000 MW electric power plant. From USAEC (1973).

isotope, $^{238}$U. Specifically, naturally occurring uranium contains 99.29 percent of $^{238}$U and only 0.71 percent of $^{235}$U (138 atoms of $^{238}$U for every atom of $^{235}$U). With a singular exception, these proportions are the same everywhere on earth because they date from the original creation of uranium by fusion and the similar decay of these isotopes since that time. The exception is a location in Oklo, Gabon, Africa, where, approximately 1.7 billion years ago, a uranium-rich mineral deposit became concentrated through sedimentation and, with the water acting as moderator (see Section 2.8.1), formed a natural nuclear reactor (Gauthier-Lafaye et al. 1996; Meshik 2005). The reactor became subcritical when water was boiled away by the reactor heat (though it restarted during subsequent flooding). The consequence was a uranium ore deposit that contained only 0.60 percent or less of $^{235}$U (as opposed to 0.71 percent elsewhere).

The nuclear fuel cycle refers to the sequence of steps in a nuclear power generation system, from the mining or acquisition of the raw ore to the refining and enrichment of that material, to its modification during power production and thence to the management of the nuclear waste. Many of the steps in a nuclear fuel cycle involve complex engineering and economics that are beyond the scope of this book (the reader could consult Knief 1992, for a comprehensive summary). However, a brief summary of commonly, envisaged fuel cycles is appropriate at this point. A basic feature of those cycles is an assay of the mass of the essential material during each step (as well as the waste). Another is the power consumption or generation during each step. One example of a nuclear fuel cycle is shown in Figure 2.1, which presents the uranium requirements for a 1000 MW pressurized water reactor.

Because $^{235}$U is the only naturally occurring fissile material, the nuclear fuel cycle must necessarily begin with the mining and milling of uranium ore. Uranium ore is relatively common, and additional recoverable resources are being discovered at a significant pace; indeed, the known resources have increased by a factor of about 3 since 1975. Some 40 percent of the known recoverable resources are found in Canada and Australia, while Russia and Kazakhstan hold another 21 percent (the highest-grade uranium ore is found in northern Saskatchewan). Thorium, an alternate nuclear reactor fuel (see Sections 2.11 and 2.2.1), is reputed to be about 3 times as abundant as uranium (WNA 2011).

Uranium is usually removed from the ore by chemical milling methods that result in the production of $U_3O_8$, known as yellowcake. The waste or *tailings* present some, primarily chemical, disposal problems. With the exception of the CANDU reactor described in Section 4.8, all other current reactors require the uranium to be *enriched*, a process in which the fraction of $^{235}U$ is increased. In preparation for enrichment, the uranium is converted to a gaseous form, namely, from $U_3O_8$ to $UF_6$, in a process known as *conversion*. Several possible methods have been used to enrich the $UF_6$, and this requires the separation of $^{235}UF_6$ from the $^{238}UF_6$, a process that cannot be accomplished chemically because these isotopes are chemically identical. The separation must therefore be accomplished physically by recourse to the small physical differences in the molecules, for example, their densities or diffusivities. The most common conversion process uses a gas centrifuge in which the heavier $^{238}UF_6$ is preferentially driven to the sides of a rapidly rotating cylinder. Another is the gaseous diffusion method, in which the gas is forced through a porous filter that the $^{235}UF_6$ penetrates more readily. In either case, a by-product is a waste known as the *enrichment tailings*.

Whether enriched or not, the fuel must then be formed into fuel ready for loading into the reactor. In most reactors this fuel fabrication stage involves conversion to solid pellets of $UO_2$ or, less commonly, UC. These cylindrical pellets are then packed into long fuel rods (as described in Section 4.3.4) whose material is referred to as *cladding*. The rods are then loaded into the reactor. The fuel cycle continues when the fuel rods are spent and removed from the reactor and the spent fuel is reprocessed.

However, before resuming this review with a description of the fuel changes that occur in a uranium reactor, it is appropriate to briefly digress to mention the other naturally available fuel, thorium, and its fuel cycle.

### 2.2.1 Thorium Fuel Cycle

The other naturally abundant element that can be used in a nuclear reactor fuel cycle is thorium, Th, whose stable isotope and fertile material is $^{232}Th$. Unlike natural uranium, natural thorium contains only trace amounts of fissile material, such as $^{231}Th$, that is insufficient to initiate a chain reaction. In a thorium-fueled reactor, $^{232}Th$ absorbs neutrons to produce $^{233}Th$ and eventually $^{233}U$ that either fissions in the reactor or is processed into new nuclear fuel. Advantages of the thorium fuel cycle include thorium's greater abundance, better physical properties and reduced plutonium production. Though thorium fuel features in a number of proposed future reactor designs (see Section 4.9.1) and in the high-temperature gas-cooled reactor (HTGR) (see Sections 2.11 and 4.6), thorium cycles are unlikely to significantly displace uranium in the nuclear power market in the near future (IAEA 2005). However, both China and India have plans for thorium cycle use in the future (Thorium Cycle Plans 2015).

### 2.2.2 Fuel Changes in the Reactor

It is appropriate to briefly review the changes in the fuel that occur during its life in the reactor core. In a typical 1000 MW light water reactor for power generation, the core contains 75,000 kg of low-enriched uranium usually in the form of $UO_2$ pellets (1000 kg of fuel typically generates about 45 kWh of electricity). During operation in a critical state, the $^{235}U$ fissions or splits producing heat in a chain reaction that also produces plutonium, other transuranic elements, and fission products. The fission fragments and heavy elements increase in concentration so that, after 18–36 months, it becomes advantageous to replace the fuel rods. At this point the fuel still contains about 96 percent of the original uranium (the term *burnup* is used to refer to the 4 percent used), but the fissionable $^{235}U$ is now less than 1 percent compared with the initial, enriched 3.5–5 percent. About 3 percent of the used fuel is waste product and 1 percent is plutonium. It is worth noting that much greater burnup (up to 20 percent) can be achieved in a fast neutron reactor (see Section 4.7).

### 2.2.3 The Postreactor Stages

Upon removal from a reactor, the fuel in the fuel rods is highly radioactive and is still producing decay heat as described in Section 2.4.2. At the time of shutdown of the reactor, the decay heat is about 6.5 percent of the full power level. This declines rapidly, falling to about 1.5 percent after an hour, 0.4 percent after a day, and 0.2 percent after a week. Spent fuel rods are therefore normally stored in isolated water pools near the generation site for several months not only to keep them cool but also to allow for the radioactive elements with short half-lives to decay substantially before further processing. The water absorbs the decay heat and prevents overheating of the fuel rods. They can be transferred to dry storage after about 5 years.

At the present time there are two subsequent strategies. The fuel may be reprocessed to recycle the useful remnants, or it may remain in long-term storage to await reevaluation of its potential use or disposal in the future. Reprocessing involves separating the uranium and plutonium from the waste products by chopping up the fuel rods (cladding and all) and dissolving them in acid to separate their components (see, e.g., Nero 1979). This enables the uranium and plutonium to be reused in fuel while the remaining 3 percent of radioactive waste is disposed of as described later. The recovered uranium is usually a little richer in $^{235}U$ than in nature and is reused after enrichment. The plutonium can be combined with uranium to make so-called *mixed oxide* (MOX) fuel that can be used as a substitute for enriched uranium in mixed oxide reactors.

All the waste from the nuclear cycle and fuel processing is classified according to the radiation it emits as either low-level, intermediate-level or high-level waste.

The high-level waste from reprocessing is reduced to powder and entombed in glass (*vitrified*) to immobilize it. The molten glass is poured into steel containers ready for long-term storage. One year of high-level waste from a 1000 MW reactor produces about 5000 kg of such high-level waste. Currently there are no disposal facilities for used fuel or for reprocessing waste. These are deposited in storage to await future use or treatment or for the creation of more permanent disposal facilities. The small mass of the material involved makes this wait not only feasible but wise.

Parenthetically, we note that the end of the Cold War created a new source of nuclear fuel from the Russian stockpiles of highly enriched weapons-grade uranium. Under a U.S.–Russian agreement, this has been diluted for use in nuclear power plants and, since then, has provided fully half of the nuclear fuel used in the United States for the generation of electricity.

## 2.3 Nuclear Physics

### 2.3.1 Basic Nuclear Fission

To proceed, it is necessary to outline the basic physics of nuclear fission. The speed of individual neutrons is quoted in terms of their kinetic energy in eV or *electron-volts*, where 1 eV is equivalent to $4.44 \times 10^{-26}$ kWh (kilowatt hours) of power. These energies range from those of so-called *fast neutrons* with energies of the order of $0.1 \rightarrow 10$ MeV down to those of so-called *thermal neutrons* with energies of the order of 0.1 eV or less. As described later, both fast and thermal neutrons play important roles in nuclear reactors.

In 1938–39 Hahn, Meitner, Strassman, and Frisch (Hahn and Strassman 1939; Meitner and Frisch 1939; Frisch 1939) first showed that any heavy atomic nucleus would undergo fission if struck by a *fast* neutron of sufficiently high kinetic energy, of the order of $1 \rightarrow 2$ MeV. Shortly thereafter, Bohr and Wheeler (1939) predicted that only very heavy nuclei containing an odd number of neutrons could be fissioned by all neutrons with kinetic energies down to the level of *thermal* neutrons (order 0.1 MeV). The only naturally occurring nucleus that meets this condition is the isotope $U^{235}$ that has 92 protons and 143 neutrons. However, the isotope $^{235}U$ is rare; in nature it only occurs as one atom for every 138 atoms of the common isotope $^{238}U$ or, in other words, as 0.71 percent of natural uranium. The consequences of this are discussed shortly.

When a neutron strikes a heavy nucleus, there are several possible consequences:

- *radiative capture* or absorption, in which the neutron is captured by the nucleus and essentially lost
- *elastic scattering*, during which the neutron rebounds from the collision without any loss of kinetic energy
- *inelastic scattering*, during which the neutron is momentarily captured and then released without fission but with considerable loss of kinetic energy
- *fission*, in which the heavy nucleus is split into several *fission fragments*, energy is generated, and several *secondary* neutrons are released

When a heavy nucleus such as $^{235}$U is fissioned by a colliding neutron, several important effects occur. First and most fundamentally for our purposes is the release of energy, mostly in the form of heat (as a result of the special theory of relativity, there is an associated loss of mass). On average, the fission of one $^{235}$U nucleus produces approximately 200 MeV ($2 \times 10^8$ eV) of energy. Thus a single fission produces roughly $8.9 \times 10^{-18}$ kWh. Because a single $^{235}$U atom weighs approximately $3.9 \times 10^{-22}$ g, it follows that the fission of one gram of $^{235}$U produces approximately 23 MWh of power. In contrast, 1 g of coal when burned produces only about $10^{-5}$ MWh, and there is a similar disparity in the waste product mass.

The second effect of a single $^{235}$U fission is that it releases two or three neutrons. In a finite volume consisting of $^{235}$U, $^{238}$U, and other materials, these so-called *prompt* neutrons can have several possible fates. They can

- collide with other $^{235}$U atoms, causing further fission
- collide with other $^{235}$U atoms and not cause fission but rather undergo radiative capture
- collide with other atoms, such as $^{238}$U, and be absorbed by radiative capture
- escape to the surroundings of the finite volume of the reactor

As a consequence, it is useful to conceive of counting the number of neutrons in a large mass in one generation and to compare this with the number of neutrons in the following generation. The ratio of these two populations is known as the *reproduction factor* or *multiplication factor*, $k$, where

$$k = \frac{\text{Number of neutrons in a generation}}{\text{Number of neutrons in the preceding generation}} \quad (2.1)$$

In addition to $k$, it is useful to define a multiplication factor that ignores the loss of neutrons to the surroundings, in other words, the multiplication factor for a reactor of the same constituents but infinite size, $k_\infty$. In the section that follows the process by which $k$ and $k_\infty$ are used in evaluating the state of a reactor is detailed.

An alternative to $k$ is the frequently used *reactivity*, $\rho$, defined as

$$\rho = \frac{(k-1)}{k} \quad (2.2)$$

and this quantity is also widely used to describe the state of a reactor. Further discussion on $k$ (or $\rho$) and $k_\infty$ and the role these parameters play in the evaluation of the criticality of a reactor is postponed until further details of the neutronics of a reactor core have been established.

### 2.3.2 Neutron Energy Spectrum

The neutrons that are released during fission have a spectrum of energies as shown in Figure 2.2, where $n(E)dE$ is the fraction of neutrons with energies in the range $E$ to $E + dE$. The distribution in Figure 2.2 is often described by empirical formulae

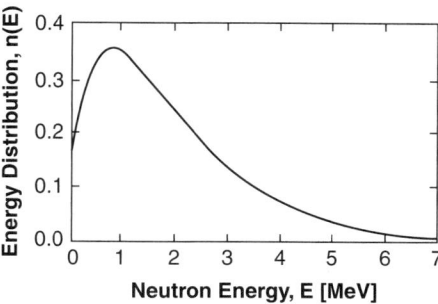

Figure 2.2. Spectrum, $n(E)$, of neutron energies due to fission.

of the type

$$n(E) = 0.453e^{-1.036E} \sinh(\sqrt{2.29E}) \tag{2.3}$$

where $E$ is in units of MeV. This integrates to unity, as it must. It follows that, as quoted earlier, the average energy of a fission neutron is 2 MeV.

### 2.3.3 Cross Sections and Mean Free Paths

In the context of nuclear interactions or events, a cross section is a measure of the probability of occurrence of that interaction or event. Consider, for example, a highly simplified situation in which $n$ neutrons per $cm^3$, all of the same velocity, $\bar{u}$ (or energy, $E$), are zooming around in a reactor volume of density $\rho$ consisting of only one type of atom of atomic weight, $A$. The number of atoms per gram is therefore $6.025 \times 10^{23}/A$, where $6.025 \times 10^{23}$ is Avagadro's number. Hence, the number of atoms per $cm^3$, $\mathcal{N}$, is given by

$$\mathcal{N} = 6.025 \times 10^{23} \rho/A \tag{2.4}$$

In such a reactor, the rate at which the moving neutrons are colliding with atoms (assumed stationary) within each $cm^3$ of volume is clearly going to be proportional to $\mathcal{N}$, to $n$, and to the velocity, $\bar{u}$, of the neutrons. The factor of proportionality, $\sigma$, or

$$\sigma = \frac{\text{Number of collisions per unit time per unit volume}}{\mathcal{N}n\bar{u}} \tag{2.5}$$

has units of area and is known as a cross section. It can be visualized as the effective frontal area of the atom that would lead to the given collision rate per unit volume (zero area would, of course, not lead to any collisions). Cross sections are measured in units called barns, where 1 barn $= 10^{-24}$ $cm^2$. They are a measure of the probability of a particular event occurring in unit volume per unit time divided by the number of collisions per unit time per unit volume, as indicated in Equation 2.5. Thus, for example, the probability of a collision causing fission is proportional to the *fission cross section*, $\sigma_f$, the probability of a collision resulting in neutron capture or absorption is proportional to the *absorption cross section*, $\sigma_a$, and the probability of a collision resulting in scattering is proportional to the *scattering cross section*, $\sigma_s$.

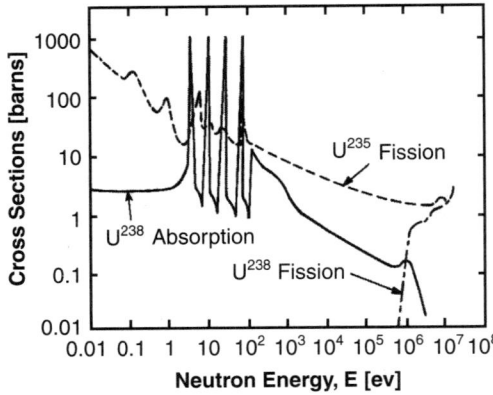

Figure 2.3. Qualitative representations of how the fission cross sections for $^{235}$U and $^{238}$U as well as the absorption cross section for $^{238}$U vary with the neutron energy.

The typical distance traveled by a neutron between such interaction events is called the mean free path, $\ell$, and this is related to the cross section as follows. Consider a given interval of time. Then the mean free path, $\ell$, will be given by the total distance traveled by all neutrons in a unit volume during that time divided by the number of neutrons undergoing a particular interaction during that time. Or

$$\ell = \frac{n\bar{u}}{\mathcal{N}n\bar{u}\sigma} = \frac{1}{\mathcal{N}\sigma} \tag{2.6}$$

More specifically, the *fission mean free path*, $\ell_f = 1/\mathcal{N}\sigma_f$, will be the typical distance traveled by a neutron between fission events, the *absorption mean free path*, $\ell_a = 1/\mathcal{N}\sigma_a$, will be the typical distance traveled by a neutron between absorption events, and so on. For this and other reasons, it is convenient to define *macroscopic cross sections*, $\Sigma$, where $\Sigma = \mathcal{N}\sigma$; these macroscopic cross sections therefore have units of inverse length.

Note that most of the cross sections, $\sigma$, that are needed for reactor analysis are strong and often complicated functions of the neutron energy, $E$. This complicates the quantitative analyses of most reactors even when the conceptual processes are quite straightforward. Qualitative examples of how some cross sections vary with $E$ are included in Figure 2.3. Rough models of how $\sigma_a$, $\sigma_f$, and $\sigma_s$ depend on $E$ (or $\bar{u}$) are as follows.

In many materials, thermal neutrons have $\sigma_a$ and $\sigma_f$ cross sections that are inversely proportional to the velocity, $\bar{u}$, and therefore vary like $E^{-\frac{1}{2}}$. In such materials it is conventional to use a factor of proportionality, $\sigma E^{\frac{1}{2}}$, at a reference state corresponding to a velocity of 2200 m/s (or an energy of $E = 0.0253$ eV). The average cross section so defined at $E = 0.0253$ eV is called the *thermal cross section reduced to 0.0253 eV* and is denoted here by $\hat{\sigma}$.

Of course, some materials like $^{238}$U have strong absorption peaks or resonances near particular energies, and, in these, the preceding model requires modification. This is often effected by supplementing the $E^{-\frac{1}{2}}$ dependence with one or more resonance peaks. Another useful observation is that scattering cross sections, $\sigma_s$, are often independent of $E$, except at high velocities, and can therefore be modeled using a single uniform value.

### 2.3.4 Delayed Neutrons and Emissions

Another important feature of nuclear fission is that although almost all of the neutrons are produced essentially instantaneously, a small fraction (about 0.7 percent) are delayed and emerge up to about 80 s after the fission event. Most of these *delayed neutrons* occur because some fission products, known as *delayed-neutron precursors*, produced by the event undergo radioactive decay and, in one or more stages of that decay, emit a neutron. One of the most common of these post fission decays occurs when the fission product $^{87}$Br decays, though there are many fission products each having several stages of decay so that delay times may range from 0.6 to 80 sec. However, for the purposes of modeling simplification, these precursors are usually divided into a small number of groups (often six) with similar properties.

These delayed neutrons play a crucial role in the control of a nuclear reactor. As described in Sections 3.9 and 4.3.6, a nuclear reactor would be very difficult to control without these delayed neutrons because a slight excess in the neutron population would grow exponentially in a matter of milliseconds (in a thermal reactor). Because of the delay times mentioned, the delayed neutrons increase this response time by several orders of magnitude and make reactor control quite manageable, as described in Section 4.3.6.

## 2.4 Radioactivity and Radioactive Decay

Additional nuclear issues that affect power generation and its auxiliary activities will now be considered. Most of these issues are related to the by-products of nuclear fission, namely, the fission products and fission radiation. Apart from the release of neutrons that was the focus of the preceding sections, nuclear fission also results in fission products, and these elements and isotopes have a number of important consequences. It also results in the emission of various types of radiation whose consequences need to be discussed. It is appropriate to begin with a brief description of radioactive decay.

### 2.4.1 Half-Life

A fundamental process that affects the behavior of a nuclear reactor and the treatment of its waste is the radioactive decay of the atomic constituents of the fuel, the fuel by-products, and the containment structures. All the heavier, naturally occurring elements of the earth and other planets were formed by fusion in the enormous thermonuclear furnace that eventually resulted in the formation of our planet and, indeed, are part of any cataclysmic astronomical event like a supernova. Only such an event could have produced the incredible temperatures (of order $10^9$ $^\circ$C) that are required for such fusion. Many of the heavier elements and isotopes formed in that event are unstable in the sense that they decay over time, fissioning into lighter elements and, at the same time, releasing radiation and/or neutrons. This release leads directly to the generation of heat through collisions (or interactions with the surrounding material) in which the kinetic energy associated with the radiation/neutrons is converted to thermal motions of the molecules of the surrounding material.

However, the rates at which these heavier elements decay differ greatly from element to element and from isotope to isotope. The rate of decay is quoted in terms of a half-life, $\tau$, namely, the length of time required for one-half of the material to be transformed and one-half to remain in its original state. Note that $\tau = 0.693/\xi$, where $\xi$ is known as the radioactive decay constant. It describes the rate of decay of the number of original nuclei of a particular isotope, $N(t)$, according to

$$-\frac{dN(t)}{dt} = \xi N(t) \tag{2.7}$$

Isotopes with extremely long half-lives, like $^{238}$U (whose half-life is $4.47 \times 10^9$ years), are therefore almost stable and, in a given period of time, exhibit very few (if any) fission events and consequently generate very little thermal energy. Rare elements with short half-lives like $^{107}$Pd may have existed at one time but have now disappeared from the earth. In between are isotopes like $^{235}$U (half-life $7.04 \times 10^8$ years) that are now much rarer than their longer-lived cousins, in this case, $^{238}$U.

## 2.4.2 Decay in a Nuclear Reactor

In the unnatural environment of a nuclear reactor, the high neutron flux causes the formation of a number of unstable isotopes. These decay to other unstable isotopes, and the chain thus followed can be long and complex before finally coming to an end with the formation of stable elements and isotopes. A full catalog of these decay chains is beyond the scope of this book, but several important examples should be given.

First note that the decay of $^{235}$U (half-life $4.47 \times 10^9$ years) results in $^{231}$Th, which, after 25.5 hours, emits radiation and becomes $^{231}$Pa. This decays with a half-life of $3.28 \times 10^4$ years to $^{227}$Ac, and the chain continues with many intermediate stages, eventually resulting in the stable lead isotope $^{207}$Pb. Many of these intermediate stages involve the series of elements with atomic numbers from 89 to 103 that are known as *actinides*. They feature prominently in the decay of a nuclear reactor and in the processing of the reactor waste.

One of the most important isotopes produced in a nuclear reactor is the unstable element plutonium, $^{239}$Pu, formed when a $^{238}$U atom absorbs a neutron. Plutonium does not occur in nature because it has a relatively short half-life ($2.44 \times 10^4$ years). Because of this short half-life, it is highly radioactive, decaying back to $^{235}$U, which then decays as described previously.

This process of decay has a number of important consequences. First, the thermal energy generated by the decay adds to the heat generated within a nuclear reactor. Thus, although the primary source of heat is the energy transmitted to the molecules of the core as a result of nuclear fission and neutron flux, the additional heat generated by decay is an important secondary contribution. This heat source is referred to as *decay heat*.

However, there is an important additional consequence for although the primary fission contribution vanishes when the reactor is shut down (when the control rods are inserted) and the neutron flux subsides, the radioactive decay continues to generate heat for some substantial time following shutdown. The decrease in heat generation occurs quite rapidly after shutdown; thus the reactor heat production decreases to 6.5 percent after 1 s, 3.3 percent after 1 min, 1.4 percent after 1 h, 0.55 percent after 1 d, and 0.023 percent after 1 yr. Though these numbers may seem small, they represent a substantial degree of heating, and coolant must be circulated through the core to prevent excessive heating that might even result in core meltdown. The production of decay heat also means that fuel rods removed from the core must be placed in a cooled environment (usually a water tank) for some time to avoid overheating.

## 2.5 Radiation

Aside from the emission of neutrons (that may be referred to as *neutron radiation*), nuclear fission and radioactive decay also result in the emission of various additional forms of radiation:

- Alpha radiation is the emission of two protons and two neutrons, an *alpha particle* being identical in composition to a nucleus of helium. It is emitted, for example, in the decay of $^{235}$U to $^{231}$Th.
- Beta radiation is the emission of small charged particles, namely, electrons and other similarly small particles. It is emitted, for example, in the decay of $^{239}$Np to $^{239}$Pu.
- Gamma radiation is the emission of short wavelength electromagnetic radiation in the form of photons. It is emitted, for example, during fission or radiative capture in $^{235}$U.

Because all of the preceding radiation emissions are associated with the decay of an isotope, a measure of radioactivity is the number of disintegrations per second, given by $\lambda N(t)$ in the notation of Equation 2.7. One disintegration per second is known as *one becquerel* (1 Bq) and is related to the more traditional unit of a *curie* (Ci) by 1 Ci $= 3.7 \times 10^{10}$ Bq. Comment on typical magnitudes of radioactivity is appropriate here. Room air has a typical radioactivity of $10^{-12}$ Ci/L or about $10^{-8}$ Ci/kg. Typical radiation treatments for cancer range up to about $10^4$ Ci, and the activity in the core of a typical thermal reactor just after it has been shut down is about $1.5 \times 10^9$ Ci.

Though they are beyond the scope of this text, the effects of radiation on materials (see, e.g., Foster and Wright 1977; Cameron 1982) or on biological tissue (see, e.g., Lewis 1977; Murray 1993) are clearly important, and therefore it is useful to establish some measures of the changes in a material or tissue brought about by exposure to radiation. These measures will clearly be a function not only of the strength and type of radiation but also of the nature of the material (or tissue) exposed to that radiation. A number of such measures are used:

- One *roentgen* (R), a traditional unit for x-rays and gamma radiation, is defined in terms of the ionization produced in air and is equivalent to the deposition in air of 87 ergs/g.
- To address the fact that the absorption of radiation in biological tissue differs from the ionization in air, the *rad* (rad) was introduced as a measure of the radiation energy absorbed per unit mass (1 rad = 100 ergs/g). One *gray* (Gr) is 100 rads and 1 Gr = 1 J/kg.
- To address the fact that the damage done depends on the type of radiation, the *roentgen-equivalent-man* (or rem for short) was introduced and defined to be the dose (energy) of 250 keV X-rays that would produce the same damage or effect as the dose (energy) of the radiation being measured. Thus 1 rem is the equivalent dose of 250 keV X-rays that would produce the same effect as 1 rad of the radiation being measured. Similarly one *sievert* (Sv) is equivalent to 1 Gr.
- The ratio of the number of rems to the number of rads is called the *quality factor*. Clearly, then, X-rays (and gamma radiation) have a quality factor of unity. In comparison, the quality factor for alpha radiation and fission fragments is 20, whereas that for neutrons varies from 5 to 20, depending on the neutron energy.

The biological effects of radiation are beyond the scope of this text (see, e.g., Lewis 1977; Murray 1993). It is sufficient for present purposes to observe that the potential damage that might be caused by the nuclear fuel before, during and after use in a nuclear reactor requires that the fuel (and other components of a reactor that may have been irradiated within the core and its immediate surroundings), be confined within a secure containment system for as long as the destructive levels of that radiation continue. Such assurance is only achieved by a system that necessarily comprises multiple systems and multiple levels of containment.

## 2.6 Containment Systems

### 2.6.1 Radioactive Release

The main safety concern with nuclear reactors has always been the possibility of an uncontrolled release of radioactive material leading to contamination and radiation exposure outside the plant. To prevent this, modern nuclear reactors incorporate three levels of containment. First, the fuel and radioactive fission products contained in the fuel pellets are packed and sealed in zirconium alloy fuel rods (see Section 4.3.4). This alloy is known by its trade name, zircaloy, and the fabrication material of the rods is known in general as *cladding*. These fuel rods, in turn, are contained inside the large, steel primary containment vessel with walls that are about 30 cm thick. The associated primary cooling piping is similarly strong. All this is then enclosed in a massive reinforced concrete structure with walls that are at least 1 m thick. Moreover, these three barriers are monitored continuously. The fuel rod walls

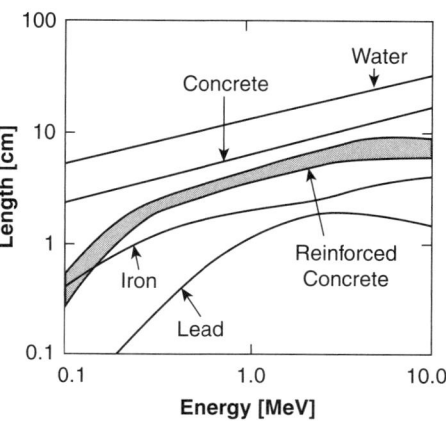

Figure 2.4. Typical distances required for a 10-fold decrease in gamma radiation in various shielding materials as a function of the energy. Adapted from Harrison (1958).

are monitored by checking for any radioactivity in the primary cooling water and that cooling system is monitored for any water leakage. Finally, the concrete structure is monitored for any air leakage.

One of these systems is the containment surrounding the operational reactor core, a barrier that is known as reactor shielding, and it is appropriate in the discussion of containment to review that topic.

### 2.6.2 Reactor Shielding

Clearly the nuclear reactor surroundings must be shielded from the intense radiation emerging from the reactor core. Man and his natural surroundings must obviously be protected from damage, but, in addition, the material of the plant must be shielded to minimize both heat damage and undesirable changes in the properties of the material, such as embrittlement (see, e.g., Foster and Wright 1977). Moreover, shielding is not only necessary for the core and the equipment enclosed in the primary reactor vessel but also for other components of the primary coolant loop, such as the pumps and heat exchangers.

In a water-cooled reactor the first level of protection is the primary cooling water surrounding the core; this water slows down the fast neutrons and provides attenuation of the gamma radiation. To supplement this, many reactor cores (including PWR cores) are surrounded by a *thermal shield*, a 3- to 7-cm-thick steel (usually stainless steel) barrel that reduces the neutron and gamma radiation impacting the inside surface of the primary pressure vessel. Incoming cooling water usually flows up the outside of the thermal shield and then down the inside before turning to flow up through the core. The steel walls of the primary pressure vessel, more than 20 cm thick, provide yet another layer of protection against the neutron and gamma radiation so that inside the concrete secondary containment structure the radiation levels are very low. That thick, reinforced concrete building ensures that the levels of radiation outside are normally very low indeed. To quantify the attenuation provided by each of these barriers, one needs to know the attenuation distances for each of the materials used and each of the proton energies. Typical data of this kind are shown in Figure 2.4.

Note also that the primary coolant water flowing through the core of a reactor carries some radioactivity out of the primary containment vessel mainly because of the radioactive nuclides, $^{16}$N and $^{19}$O, formed when water is irradiated. These isotopes, $^{16}$N and $^{19}$O, have half-lives of only 7 s and 29 s, respectively, although they produce gamma radiation during decay (Gregg King 1964). Thus, for example, access to secondary containment structures is restricted during reactor operation.

## 2.7 Natural Uranium Reactors

A useful and appropriate starting point for the discussion of nuclear reactors is to consider the state of naturally occurring uranium. As noted earlier, the most common isotope is $^{238}$U, and the fission cross section for $^{238}$U has the form shown in Figure 2.3. Thus only high-energy or *fast neutrons* with energies greater than about 2 MeV can cause fission of $^{238}$U. However, the absorption and scattering cross sections are much larger, and therefore any population of neutrons in $^{238}$U rapidly declines; such a *reactor* is very subcritical.

Now consider a naturally occurring mixture of $^{238}$U and $^{235}$U. As previously stated and illustrated in Figure 2.3, the "fissile" isotope $^{235}$U can be fissioned even with low-energy neutrons, and therefore the presence of the $^{235}$U causes an increase in the reactivity of the mixture. However, the high absorption cross section of the "fertile" isotope $^{238}$U still means that the reactivity of the mixture is negative. Thus no chain reaction is possible in natural uranium. One can visualize that if it were possible, then this would have happened at some earlier time in the earth's evolution and that no such unstable states or mixtures could be left today. Such has also been the fate of higher-atomic-weight elements that may have been produced during nuclear activity in the past.

There are several different ways in which the naturally occurring uranium mixture might be modified to produce a critical or supercritical chain reaction in which the neutron population is maintained. One obvious way is to create a mixture with a higher content of $^{235}$U than occurs naturally. This is called enriched uranium and requires a process of separating $^{238}$U and $^{235}$U in order to generate the enriched mixture. Because $^{238}$U and $^{235}$U are almost identical chemically and physically, separation is a difficult and laborious process, the main hurdle during the Manhattan project.

## 2.8 Thermal Reactors

### 2.8.1 Moderator

An alternative that avoids the costly and difficult enrichment process and eliminates the need to handle weapons-grade uranium is hinged on a characteristic of the absorption cross section of $^{238}$U whose form was shown in Figure 2.3. This has strong peaks at intermediate neutron energies, the so-called *capture resonances*, so that many neutrons, slowed down by scattering, are absorbed by the $^{238}$U before they can reach low or *thermal* energies. This is important because, as shown in Figure 2.3,

$^{235}$U has a very high fission cross section at thermal energies, and this potential source of fast neutrons is attenuated because so few neutrons can pass through the resonance barrier. Note that neutrons that are in the process of being slowed down are termed *epithermal neutrons*.

However, if it were possible to remove the fast neutrons from the reactor, slow them down to thermal energies, and then reintroduce them to the core, the reactivity of the reactor could be increased to critical or supercritical levels. In practice, this can be done by including in the reactor a substance that slows down the neutrons without absorbing them. These slowed-down neutrons then diffuse back into the uranium and thus perpetuate the chain reaction. This special substance is known as the *moderator*, and it transpires that both water and carbon make good moderators. Such a reactor is called a *thermal* reactor because its criticality is heavily dependent on the flux of low-energy, thermal neutrons. Virtually all the nuclear reactors used today for power generation are thermal reactors, and this monograph therefore emphasizes this type of reactor.

To summarize, a conventional thermal reactor core comprises the following components:

- natural or slightly enriched uranium fuel, usually in the form of an oxide and encased in fuel rods to prevent the escape of dangerous fission products
- moderator, usually water (sometimes heavy water) or carbon
- control rods made of material that is highly absorbent of neutrons so that the insertion or withdrawal of the rods can be used to control the reactivity of the core
- a cooling system to remove the heat, the energy produced; in many reactors, water serves as both the coolant and the moderator

A variety of thermal reactors have been developed and used to produce power in the world. These comprise three basic types:

1. Light water reactors (LWRs) are by far the most common type used for power generation and include the common pressurized water reactors (PWRs) and boiling water reactors (BWRs) (see Sections 4.3.1 and 4.3.3). They use regular water (so-called *light* water) as both the coolant and the moderator but need somewhat enriched uranium fuel (about 2 percent $^{235}$U).
2. Heavy water reactors (HWRs) use natural, unenriched uranium fuel and achieve the needed increase in reactivity by using deuterium oxide (*heavy* water) as the moderator and coolant rather than light water. The Canadian CANDU reactor is the best-known example of this type.
3. There are gas-cooled reactors (GCRs) in which the primary coolant loop utilizes a gas (e.g. carbon dioxide or helium) rather than water. Typically these use graphite as the moderator. Examples are the high-temperature gas-cooled reactor (HTGR) and the advanced gas-cooled reactor (AGR) manufactured respectively in the United States and the United Kingdom.

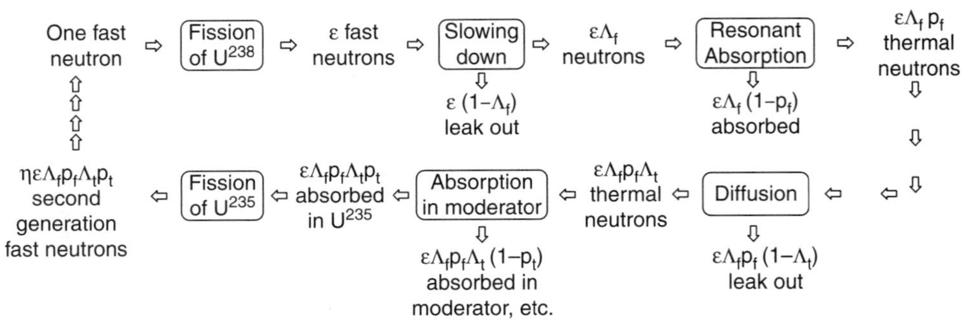

One fast neutron ⇒ [Fission of U²³⁸] ⇒ ε fast neutrons ⇒ [Slowing down] ⇒ εΛ_f neutrons ⇒ [Resonant Absorption] ⇒ εΛ_f p_f thermal neutrons

ε(1−Λ_f) leak out

εΛ_f(1−p_f) absorbed

ηεΛ_f p_f Λ_t p_t second generation fast neutrons ⇐ [Fission of U²³⁵] ⇐ εΛ_f p_f Λ_t p_t absorbed in U²³⁵ ⇐ [Absorption in moderator] ⇐ εΛ_f p_f Λ_t thermal neutrons ⇐ [Diffusion]

εΛ_f p_f Λ_t(1−p_t) absorbed in moderator, etc.

εΛ_f p_f(1−Λ_t) leak out

Figure 2.5. Simplified history of neutrons in a thermal reactor.

This list focuses on the large thermal reactors for power generation. There is a much greater variety of design in the smaller reactors used for research and for power sources in vehicles such as submarines and space probes. The various types of thermal reactors are examined in more detail in Chapter 4.

## 2.8.2 Neutron History in a Thermal Reactor

Figure 2.5 delineates the typical neutron history in a thermal reactor. Using a single fast neutron as an arbitrary starting point (upper left), this fast neutron fissions a $^{238}$U atom and produces $\epsilon$ fast neutrons. Some fraction, $(1 - \Lambda_F)$, of these fast neutrons leaks out through the boundaries of the reactor and another fraction, $(1 - P_F)$, is absorbed in $^{238}$U leaving $\epsilon \Lambda_F P_F$ that have been slowed down to thermal speed either in the moderator or otherwise. Some fraction, $(1 - \Lambda_T)$, of these thermal neutrons also leaks out through the boundaries and another fraction, $(1 - P_T)$, is absorbed in the $^{238}$U or the moderator or other material. This finally leaves $\epsilon \Lambda_F P_F \Lambda_T P_T$ thermal neutrons to cause fission of $^{235}$U and thus produce $\eta \epsilon \Lambda_F P_F \Lambda_T P_T$ second-generation fast neutrons. In this history, $\eta$ is the *thermal fission factor of* $^{235}$U, $\epsilon$ is the *fast fission factor*, $\Lambda_F$ is the *fast neutron nonleakage probability*, $\Lambda_T$ is the *thermal neutron nonleakage probability*, $P_F$ is the *resonance escape probability*, and $P_T$ is the *thermal utilization factor for* $^{235}$U.

It follows that the multiplication factors, $k$ and $k_\infty$, are given by

$$k = \eta \epsilon \Lambda_F P_F \Lambda_T P_T \qquad k_\infty = \eta \epsilon P_F P_T \qquad (2.8)$$

known respectively as the *six-factor formula* and the *four-factor formula*. It also follows that a reactor operating at steady state will have $k = \eta \epsilon \Lambda_F P_F \Lambda_T P_T = 1$ and the control system needed to maintain such steady state operation must be capable of adjusting one or more of the factors $P_F$ and $P_T$.

The thermal energy resulting from this process comes mostly from the fission process, and therefore both the neutron population and the neutron flux (see Section 3.2) are roughly proportional to the rate of generation of heat within a reactor core. Thus an evaluation of the neutron flux by the methods of Chapter 3 can be used to estimate the generation of heat within the components of the core, as described in Chapter 5.

## 2.9 Fast Reactors

An alternative to the thermal reactor strategy is to strive to attain criticality using, primarily, fast neutrons. Various fuels and combinations of fuels can provide the required self-sustaining reaction. Highly enriched uranium (over 20 percent $^{235}$U) is possible, and in this process, $^{238}$U produces several isotopes of plutonium including $^{239}$Pu and $^{241}$Pu by neutron capture. Then the $^{239}$Pu and $^{241}$Pu undergo fission and produce heat in the same way as $^{235}$U or $^{233}$U. The $^{238}$U is referred to as the *fertile* material while, like $^{235}$U or $^{233}$U, the $^{239}$Pu and $^{241}$Pu are referred to as *fissile* materials. An alternative is the fertile thorium, $^{232}$Th, that yields fissile thorium.

Although fast reactors could use enriched uranium, they are more efficiently fueled with fissile plutonium or a mixture of uranium and plutonium. In the latter case the $^{238}$U will produce more plutonium. A reactor in which the net change of plutonium content is negative is called a *burner* fast reactor, whereas a reactor in which the plutonium content is increasing is termed a *fast breeder reactor* (FBR). Commonly, fast breeder reactors are cooled using liquid metal (sodium, lead, or mercury) rather than water and so are referred to as *liquid metal fast breeder reactors* (LMFBR). Almost all the commercial fast reactors constructed to date are LMFBRs, hence the focus on this type in the pages that follow.

The advantage of a fast reactor is that it makes much better use of the basic uranium fuel, indeed, by an estimated factor of 60. Moreover, because an FBR breeds new fuel, there are subsequent savings in fuel costs because the spent fuel can be reprocessed to recover the usable plutonium. Examples of LMFBRs are the French-built Phenix (and Superphenix) and the Russian BN-600 reactor that has been generating electricity since 1980 (see Section 4.8). However, as described in Section 7.6.3, the safety issues associated with these fast reactors are much more complex than are those with thermal reactors.

## 2.10 Criticality

The discussion of the criticality of a nuclear reactor is be resumed. It is self-evident that a finite reactor will manifest an accelerating chain reaction when $k > 1$ (or $\rho > 0$); such a reactor is termed *supercritical*. Moreover a reactor for which $k = 1$ ($\rho = 0$) is termed *critical* and one for which $k < 1$ ($\rho < 0$) is *subcritical*. Note that because the neutron escape from a finite reactor of typical linear dimension, $l$, is proportional to the surface area, $l^2$, while the neutron population and production rate will be proportional to the volume, $l^3$, it follows that $k$ will increase linearly with the size, $l$, of the reactor and hence there is some *critical size* at which the reactor will become critical. It is clear that a power plant needs to maintain $k = 1$ to produce a relatively stable output of energy while gradually consuming its nuclear fuel.

Consequently, two sets of data determine the criticality of a reactor. First, there is the basic neutronic data (the fission, scattering, and absorption cross sections, and other details that are described previously in this chapter); these data are functions of the state of the fuel and other constituents of the reactor core but are independent of

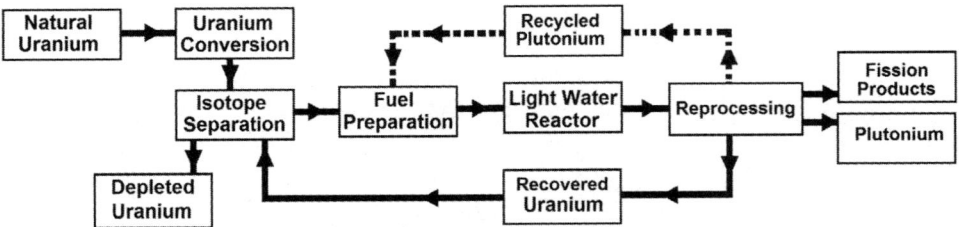

Figure 2.6. The conventional light water reactor (LWR) fuel cycle (solid line) with plutonium recycling (dashed line) and without (no dashed line).

the core size. These so-called *material properties* of a reactor allow evaluation of $k_\infty$. The second set of data is the geometry of the reactor that determines the fractional leakage of neutrons out of the reactor. This is referred to as the *geometric property* of a reactor, and this helps define the difference between $k$ and $k_\infty$. These two sets of data are embodied in two parameters called the *material buckling*, $B_m^2$, and the *geometric buckling*, $B_g^2$, that are used in evaluating the criticality of a reactor. These will be explicitly introduced and discussed in Chapter 3.

## 2.11  Fuel Cycle Variations

To conclude the discussion of nuclear fuel cycles, it is appropriate to reprise the variations in the fuel cycle represented by the present family of nuclear power generating reactors. The basic fuel cycle for a light water reactor (LWR) (see Section 4.3.1) is depicted in Figure 2.6, but without the dashed line indicating plutonium recycling. As described, the basic cycle begins with enriched uranium (3.5–5 percent $^{235}$U as compared to the 0.71 percent in natural uranium). The depleted uranium from the fuel preparation process contains about 0.2 percent $^{235}$U. Spent fuel removed from the reactor contains about 0.8 percent $^{235}$U and the fission products described previously, as well as plutonium. As indicated by the dashed line in Figure 2.6, the plutonium can be recycled and used again in a fuel in which it is mixed with uranium that might typically only need to be enriched to about 2.0 percent $^{235}$U. Such a *mixed oxide fuel* (MOX) consisting of $UO_2$ and $PuO_2$ needs to be carefully adjusted to have the desired neutronic activity.

As a second example, note the very different fuel cycle for the high-temperature gas reactor, HTGR (see Section 4.6), depicted in Figure 2.7, that utilizes thorium as

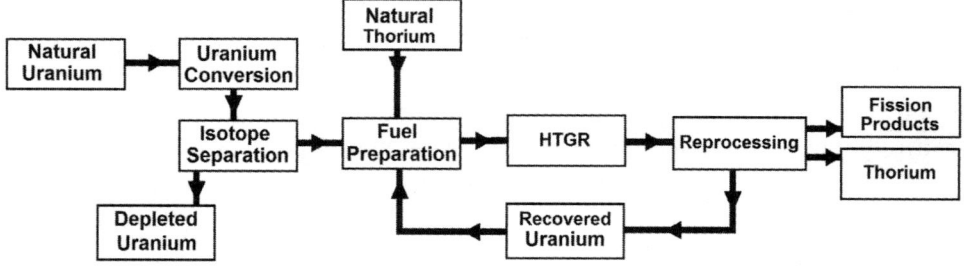

Figure 2.7. The thorium fuel cycle for the high-temperature gas reactor (HGTR).

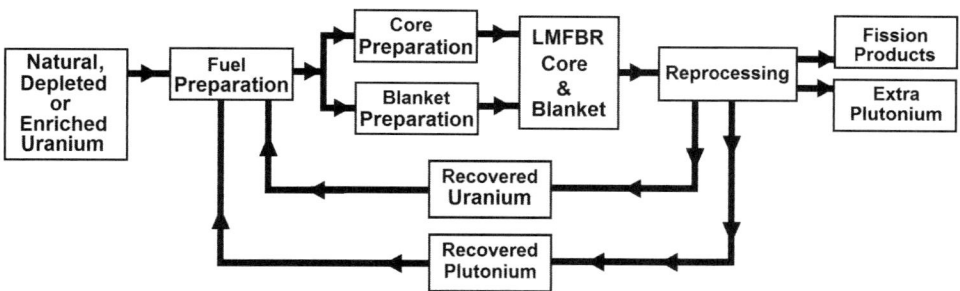

Figure 2.8. The liquid metal fast breeder reactor (LMFBR) fuel cycle.

the primary fertile material. This is mixed with highly enriched uranium (93 percent $^{235}$U) to provide the necessary neutron activity. In the reactor the thorium produces $^{233}$U that can then be recycled in mixed fuel, MOX.

As a third example, the fuel cycle for a typical liquid metal fast breeder reactor, LMFBR (see Section 4.8), is shown in Figure 2.8. This may be fueled with a mix of natural, depleted (recycled), or enriched uranium as well as recycled plutonium. As described in Sections 4.7 and 4.8, the driver core of an LMFBR is surrounded by a blanket in which natural uranium produces plutonium that can later be recycled in new fuel. This recycling of plutonium (as well as uranium) makes much more thorough and efficient use of the basic uranium fuel and therefore not only extends the potential use of the natural uranium resource but also reduces the cost of the power produced.

Finally, it should be noted that there is significant potential for interactions between the various fuel cycles. These interactions allow for increased efficiency in the utilization of the limited natural resources and also allow improved cost effectiveness. Moreover, the potential for the development of improved fuel cycles in the future means that temporary or retrievable storage of nuclear waste may be the optimum strategy.

## REFERENCES

Bohr, N., and Wheeler, J. A. (1939). The mechanism of nuclear fission. *Physical Review*, **56**, 426–450.

Cameron, I. R. (1982). *Nuclear fission reactors*. Plenum Press.

Foster, A. R., and Wright, R. L., Jr. (1977). *Basic nuclear engineering*. Allyn and Bacon.

Frisch, O. R. (1939). Physical evidence for the division of heavy nuclei under neutron bombardment. *Nature*, **143**, 276–278.

Gauthier-Lafaye, F., Holliger, P., and Blanc, P.-L. (1996). Natural fission reactors in the Franceville Basin, Gabon: A review of the conditions and results of a "critical event" in a geologic system. *Geochimica et Cosmochimica Acta*, **60**, 4831–52.

Gregg King, C. D. (1964). *Nuclear power systems*. Macmillan.

Hahn, O., and Strassmann, F. (1939). On the detection and characteristics of the alkaline earth metals formed by irradiation of uranium with neutrons. *Die Naturwissenschaften*, **27**, 11–15.

Harrison, J. R. (1958). *Nuclear reactor shielding*. Temple.

IAEC. (2005). *Thorium fuel cycle: potential benefits and challenges*. International Atomic Energy Agency, IAEA-TECDOC-1450.

Knief, R. A. (1992). *Nuclear engineering: Theory and practice of commercial nuclear power.* Hemisphere.

Lewis, E. E. (1977). *Nuclear power reactor safety.* John Wiley.

Meitner, L., and Frisch, O. R. (1939). Disintegration of uranium by neutrons: A new type of nuclear reaction. *Nature*, **143**, 239–40.

Meshik, A. P. (2005). *The workings of an ancient nuclear reactor.* Scientific American.

Murray, R. L. (1993). *Nuclear energy.* Pergamon Press.

Nero, A. V. (1979). *A guidebook to nuclear reactors.* University of California Press.

Thorium Cycle Plans. (2015). http://www.kalpakkam.com/node/294.

USAEC. (1973). *The nuclear industry.* US Atomic Energy Commission Rep. WASH-1174-73.

WNA (2011). *World Nuclear Association Information Library.* http://www.world-nuclear.org/Information-Library/.

# 3

## Core Neutronics

### 3.1 Introduction

To analyze the power generated in the reactor core and thus the temperature distribution, one must first calculate the neutron population distribution and neutron flux and examine how they vary with the control devices. This chapter describes how this can be done.

### 3.2 Neutron Density and Neutron Flux

It is convenient to begin by defining several characteristic features of neutron transport and by introducing the concept of neutron density, $N$, a measure of the number of free neutrons per unit volume. Of course, this may be a function of time, $t$, and of position, $x_i$, within the core. Furthermore, these neutrons may have a range of different energies, $E$, and the number traveling in a particular angular direction, $\Omega_j$ (a unit vector), may have a different density than those traveling in another direction. Consequently, to fully describe the neutron density, $N$ must be considered to be a function of $x_i, t, E$, and $\Omega_j$, and the number of neutrons in a differential volume $dV$ that have energies between $E$ and $E + dE$ and are traveling within the small solid angle, $d\Omega$, around the direction $\Omega_j$ would be

$$N(x_i, t, E, \Omega_j) \, dV \, dE \, d\Omega \qquad (3.1)$$

Consequently, $N$ has units of number per unit volume per unit energy per unit solid angle. Even for a simple core geometry, $N$, when discretized, is a huge matrix, especially because the energy spectrum may require very fine discretization to accurately portray the variation with $E$ (see, e.g., Figure 2.3).

Denoting the magnitude of the neutron velocity by $\bar{u}(x_i, t, E)$ (a function of position, time, and energy but assumed independent of direction, $\Omega_j$), it is conventional to define the *angular neutron flux*, $\varphi$, by

$$\varphi(x_i, t, E, \Omega_j) = N(x_i, t, E, \Omega_j) \, \bar{u}(E) \qquad (3.2)$$

The conventional semantics here are somewhat misleading because $\varphi$ is not a flux in the sense that that term is commonly used in physics (indeed, $\varphi$ as defined earlier is a scalar, whereas a conventional flux is a vector); it is perhaps best to regard $\varphi$ as a convenient mathematical variable whose usefulness will become apparent later.

A more physically recognizable characteristic is the conventional vector quantity known as the *angular current density*, $J_j^*$, given by

$$J_j^*(x_i, t, E, \Omega_j) = \bar{u}\,\Omega_j\, N(x_i, t, E, \Omega_j) = \Omega_j\, \varphi(x_i, t, E, \Omega_j) \qquad (3.3)$$

because $\bar{u}\Omega_j$ is the vector velocity of a neutron traveling in the direction $\Omega_j$. This angular current density, $J_j^*$, might be more properly called the neutron flux, but confusion would result from altering the standard semantics. The physical interpretation is that $J_j^* dE d\Omega$ is the number of neutrons (with energies between $E$ and $E + dE$) traveling within the solid angle $d\Omega$ about the direction $\Omega_j$ per unit area normal to that direction per unit time. Note that because $\Omega_j$ is a unit vector, the magnitude of $J_j^*$ is $\varphi$.

The preceding definitions allow for the fundamental quantities $\varphi$ and $J_j^*$ to vary with the angular orientation $\Omega_j$. However, it will often be assumed that these variations with orientation are small or negligible. Then integration over all orientations allows the definition of an angle-integrated neutron flux, $\phi(x_i, t, E)$ (later abbreviated to *neutron flux*), and an angle-integrated current density, $J_j(x_i, t, E)$:

$$\phi(x_i, t, E) = \int_{4\pi} \varphi(x_i, t, E, \Omega_j)d\Omega \qquad (3.4)$$

$$J_j(x_i, t, E) = \int_{4\pi} J_k^*(x_i, t, E, \Omega_j)d\Omega \qquad (3.5)$$

Note that if $\varphi(x_i, t, E, \Omega_j)$ and/or $J_j^*(x_i, t, E, \Omega_j)$ are isotropic and therefore independent of $\Omega_j$, then

$$\phi = 4\pi\varphi \qquad J_j = 4\pi J_j^* \qquad (3.6)$$

In the simpler neutronics calculations later in this book, the *neutron flux*, $\phi$, is the dependent variable normally used in the calculations.

## 3.3 Discretizing the Energy or Speed Range

To calculate the neutron distribution in every detail, one would need to consider the population of neutrons with a particular energy and direction of motion at every location and at every moment in time and be able to analyze their collisions, production, and capture. This is an enormous computational challenge, particularly because the cross sections for those interactions are all complicated functions of the neutron energy. The problem is further complicated by the fact that the mean free paths are comparable with the dimensions of the detailed interior structure of the reactor core (e.g., the fuel rod diameter or coolant channel width). The general approach to this problem is known as *neutron transport theory*. The details of the general theory, for which the reader is referred to other classic texts such as Glasstone and Sesonske (1981) or Duderstadt and Hamilton (1976), are beyond the scope of this monograph. In part, this is because most practical calculations are performed only after radical simplifications that are necessary to arrive at a practical computation of the neutron dynamics in a practical reactor.

Before further deliberation of neutron transport theory, some of the approximations that will be made later in the analysis can be anticipated. As implied in the preceding section, the neutron energies represented in a reactor cover a wide range of speeds, and, because each speed may have different cross sections to various reactions, it becomes extremely complicated to incorporate all of these intricate details. Fortunately, it is sufficient for many purposes to discretize the energy range in very crude ways. The crudest approach is to assume that all the neutrons have the same energy: a thermal energy in thermal reactors because most of the heat produced is generated by fission that is proportional to the thermal neutron flux. This approach is further pursued in Section 3.6.3.

One of the first hurdles experienced in implementing a method with a very crude discretization of the energy spectrum is the need to find average cross sections that are applicable to the assumed, uniform energy within each subrange. This can be effected by using the *reduced thermal models* described in Section 2.3.3. Thus a one-speed thermal neutron model could have a single neutron energy of $E = 0.0253$ eV and an absorption cross section of $\hat{\sigma}$. If the corresponding thermal neutron flux (called the *flux reduced to* 0.0253 eV) is also denoted by a hat, or $\hat{\phi}$, then the rate of absorption would be given by $\mathcal{N}\hat{\phi}\hat{\sigma}$. Henceforth, this averaging will be adopted and, for the sake of simplicity, the hat will be omitted and $\sigma$ and $\phi$ will be used to denote the averaged cross section and the averaged neutron flux.

## 3.4 Averaging over Material Components

The preceding averaging referred, of course, to the process of averaging within one (or sometimes two) range(s) of neutron energy within a given material. However, a reactor core consists of many different physical components, each of which may have different absorption and scattering properties. Thus, in addition to the energy averaging described earlier, the simplest models homogenize this core by also averaging over these physical components, as follows. The total reaction or absorption rate (per unit total core volume) in the homogenized core is clearly the sum of the reaction rates in each of the $M$ materials present (denoted by the superscript $m = 1$ to $m = M$). The reaction rate in the material $m$ per unit total core volume will be given by $\mathcal{N}^m \alpha^m \sigma^m \phi^m = \alpha^m \Sigma^m \phi^m$, where $\mathcal{N}^m$ is the number atoms of material $m$ per unit volume of $m$, $\alpha^m$ is the volume of $m$ per unit total core volume, $\sigma^m$ is the reaction cross section for the material $m$, $\Sigma^m$ is the corresponding macroscopic cross section (see Section 2.3.3), and $\phi^m$ is the neutron flux in the material $m$. It follows that the average neutron flux, $\phi$, and the average macroscopic absorption cross section, $\Sigma$, will be related by

$$\Sigma\phi = \sum_{m=1}^{M} \mathcal{N}^m \alpha^m \sigma^m \phi^m \qquad (3.7)$$

Note that in the special case in which the typical physical dimensions of the components are much smaller than the neutron mean free path, the neutron flux should be considered identical in all the materials ($\phi^m = \phi$) so that

$$\Sigma = \sum_{m=1}^{M} \mathcal{N}^m \alpha^m \sigma^m = \sum_{m=1}^{M} \alpha^m \Sigma^m \qquad (3.8)$$

This allows evaluation of the effective cross sections for a core with physically different components. Of course, each interaction or event will have its own effective cross section so that there will be cross sections for fission, $\Sigma_f$, for absorption, $\Sigma_a$, for scattering, $\Sigma_s$, and so on.

### 3.5 Neutron Transport Theory

The first simplification of *neutron transport theory* is to assume that the range of neutron energies can be discretized into a small number of energy ranges (sometimes, as has been described in the preceding section, the even more radical assumption is made that all neutrons have the same energy). Then the heart of neutron transport theory is a neutron continuity equation known as the *neutron transport equation* that simply represents the neutron gains and losses for an arbitrary control volume, $V$, within the reactor for each of the ranges of neutron energies being considered. In evaluating this neutron balance for each of the energy ranges, it is necessary to account for

[A] the rate of increase of those neutrons within the volume $V$
[B] the rate of appearance of those neutrons in $V$ as a result of flux through the surface of the volume $V$
[C] the loss of those neutrons as a result of absorption (and as a result of scattering to an energy level outside of the entire range of discretized energies)
[D] the rate of appearance of those neutrons that, as a result of a scattering interaction, now have energies of the magnitude being evaluated
[E] the rate of production of those neutrons in $V$, most importantly by fission

These alphabetical labels will be retained when each of these individual terms is considered in the analysis that follows.

   The second simplification, mentioned earlier, recognizes that the angular variations in the neutron flux are rarely of first-order importance. Hence nonisotropic details can be laid aside and the neutron flux can be integrated over the angular orientation, $\Omega_j$, as described in Equations 3.4 and 3.5. When this integration is performed on the neutron transport equation to extract an equation for the integrated neutron flux, $\phi(x_i, t, E)$, the result takes the following form (Glasstone and Sesonske 1981; Duderstadt and Hamilton 1976):

$$\frac{1}{\bar{u}} \frac{\partial \phi}{\partial t} + \frac{\partial J_j}{\partial x_j} + \Sigma_a \phi = \int_0^\infty \Sigma_s(E' \rightarrow E)\phi(x_i, t, E')dE' + S(x_i, t, E) \qquad (3.9)$$

This is known as the *neutron continuity equation*. The five terms each represent a contribution to the population (per unit volume) of neutrons of energy $E$ at the location $x_i$ and the time $t$, specifically,

[A] The first term is the rate of increase of neutrons in that unit volume.
[B] The second term is the flux of neutrons out of that unit volume.
[C] The third term is the rate of loss of neutrons due to absorption.
[D] The fourth term is the rate of increase of neutrons of energy $E$ due to scattering where the energy before the scattering interaction was $E'$. Consequently, an integration over all possible previous energies, $E'$, must be performed.
[E] The fifth term is the rate of production of neutrons of energy $E$ within the unit volume due to fission, $S(x_i, t, E)$.

Consequently, the following nomenclature pertains in Equation 3.9: $\phi(x_i, t, E)$ and $J_j(x_i, t, E)$ are the angle-integrated flux and current density as defined by Equations 3.4 and 3.5, $\bar{u}$ represents the magnitude of the neutron velocity (assumed isotropic), $\Sigma_a(x_i, E)$ is the macroscopic cross section at location $x_i$ for collisions in which neutrons of energy, $E$, are absorbed, $\Sigma_s(E' \rightarrow E)$ is the macroscopic cross section for scattering of neutrons of energy $E'$ to energy $E$, and $S(x_i, t, E)$ is the rate of production in a unit volume at $x_i$ and $t$ of neutrons of energy $E$.

Assuming that the macroscopic cross sections and the source term are given, Equation 3.9 is the equation that determines the population of neutrons for each energy level $E$ as a function of position $x_i$ and time $t$. Ideally this equation should be solved for the neutron flux, $\phi(x_i, t, E)$. However, there remains a problem in that the equation involves two unknown functions, $\phi(x_i, t, E)$ and $J_j(x_i, t, E)$, a problem that was further complicated by the integration over the angle. Specifically, whereas $\varphi(x_i, t, E, \Omega_j)$ and $J_j^*(x_i, t, E, \Omega_j)$ are simply related by Equation 3.3, the functions, $\phi(x_i, t, E)$ and $J_j(x_i, t, E)$, defined respectively by Equations 3.4 and 3.5, are not so easily related.

To proceed with a solution, another relation between $\phi(x_i, t, E)$ and $J_j(x_i, t, E)$ must be found. One simple way forward is to heuristically argue that in many transport processes (e.g., the conduction of heat), the concentration (in this case $\phi$) and the flux (in this case $J_j$) are simply connected by a relation known as Fick's law, in which the flux is proportional to the gradient of the concentration, the factor of proportionality being a diffusion coefficient. This assumption or approximation is made here by heuristically declaring that

$$J_j(x_i, t, E) = -D(x_i)\frac{\partial \phi(x_i, t, E)}{\partial x_j} \qquad (3.10)$$

where $D$ is a diffusion coefficient that may be a function of position. This diffusive process could be viewed as the effective consequence of neutrons undergoing multiple scattering interactions, just as heat diffusion is the effective consequence of molecules undergoing multiple interactions. One of the modern computational approaches to neutron transport, known as the Monte Carlo method (see Section 3.11), utilizes this general consequence.

Fick's law will be the model that will be the focus here. However, it is valuable to point out that Fick's law for neutrons can also be derived from the basic conservation laws in the following way. Returning to the neutron continuity principle, one can propose an expansion for the neutron flux, $\varphi$, that includes the angle-integrated average used previously plus a perturbation term that is linear in the angle $\Omega_j$. Assuming that this second term is small (that the flux is only weakly dependent on the angle), one can then establish the equation for this linear perturbation term that emerges from the neutron continuity principle. Making some further assumptions (neglect of the time-dependent term, assumption of isotropic source term), the result that emerges from this perturbation analysis is

$$\frac{1}{3}\frac{\partial \phi}{\partial x_j} + \Sigma_{tr} J_j = 0 \tag{3.11}$$

where $\Sigma_{tr}$ is called the *macroscopic transport cross section* and is given by $\Sigma_{tr} = \Sigma_a + \Sigma_s - \mu \Sigma_s$, where $\mu$ is the cosine of the average scattering angle. (For further detail and a rigorous derivation of these relations, the reader should consult texts such as Glasstone and Sesonske 1981; Duderstadt and Hamilton 1976). Comparing Equation 3.11 with Equation 3.10, it can be observed that $J_j$ and $\phi$ do, indeed, connect via Fick's law and that the *neutron diffusion coefficient, $D(x_i)$*, is given by

$$D(x_i) = \frac{1}{3\Sigma_{tr}} \tag{3.12}$$

Equation 3.11 can then be used to substitute for $J_j$ in Equation 3.9 and thus generate an equation for the single unknown function, $\phi(x_i, t, E)$.

Computational methods based on the assumption of Equation 3.10 are known as diffusion theories, and these are the focus of the sections that follow.

### 3.6 Diffusion Theory

#### 3.6.1 Introduction

It is appropriate to recall at this point that diffusion theory for the neutronics of a reactor core avoids much complexity posed by the interior structure of the reactor core by assuming the following:

1. The reactor core can be considered to be homogeneous. As described in Section 3.3, this requires the assumption that the neutron mean free paths are long compared with the typical small-scale interior dimensions of the reactor core (such as the fuel rod dimensions). This then allows characterization of the dynamics by

a single neutron flux, $\phi$, though one that varies with time and from place to place. Fuel rods are typically only a few centimeters in diameter, and with neutron diffusion lengths, $L$ (see Equation 3.19), of about 60 cm this criterion is crudely satisfied in most thermal reactors.

2. The characteristic neutron flux does not vary substantially over one mean free path. This is known as a *weakly absorbing* medium.
3. The reactor core is large compared with the neutron mean free paths so that a neutron will generally experience many interactions within the core before encountering one of the core boundaries. Most thermal reactor cores are only a few neutron diffusion lengths, $L$, in typical dimension, so this criterion is only very crudely satisfied.

These last two assumptions effectively mean that neutrons diffuse within the core and the overall population variations can be characterized by a diffusion equation.

In addition to the governing equation, it is necessary to establish both initial conditions and boundary conditions on the neutron flux, $\phi$. Initial conditions will be simply given by some known neutron flux, $\phi(x_i, 0)$, at the initial time, $t = 0$. The evaluation of boundary conditions requires the development of relations for the one-way flux of neutrons through a surface or discontinuity. To establish such relations, the one-way flux of neutrons through any surface or boundary (the coordinate $x_n$ is defined as normal to this boundary in the positive direction) will be denoted by $J_n^+$ in the positive direction and by $J_n^-$ in the negative direction. Clearly the net flux of neutrons will be equal to $J_n$ so that

$$J_n^+ - J_n^- = J_n = -D \frac{\partial \phi}{\partial x_n} \tag{3.13}$$

using Equations 3.11 and 3.12. Conversely, the sum of these same two fluxes must be related to the neutron flux, specifically,

$$J_n^+ + J_n^- = \frac{1}{2} \phi \tag{3.14}$$

(see, e.g., Glasstone and Sesonske 1981). The factor of one-half is geometric: because the flux $\phi$ is in all directions, the resultant in the direction $x_n$ requires the average value of the cosine of the angle relative to $x_n$.

It follows from Equations 3.13 and 3.14 that

$$J_n^+ = \frac{1}{4} \phi - \frac{D}{2} \frac{\partial \phi}{\partial x_n} \qquad J_n^- = \frac{1}{4} \phi + \frac{D}{2} \frac{\partial \phi}{\partial x_n} \tag{3.15}$$

These relations allow the establishment of boundary conditions when the condition involves some constraint on the neutron flux. Two examples will suffice.

At an interface between two different media denoted by subscripts 1 and 2 (and with diffusion coefficients $D_1$ and $D_2$), the neutron flux into medium 1 must be equal to the neutron flux out of medium 2 and, conversely, the neutron flux out of medium

1 must be equal to the neutron flux into medium 2. Therefore, from Equations 3.15, it follows that, at the interface,

$$\phi_1 = \phi_2 \qquad D_1 \frac{\partial \phi_1}{\partial x_n} = D_2 \frac{\partial \phi_2}{\partial x_n} \qquad (3.16)$$

A second, practical example is the boundary between one medium (subscript 1) and a vacuum from which there will be no neutron flux back into the first medium. This is an approximation to the condition at the surface boundary of a reactor. Then, on that boundary, it is clear that $J_n^- = 0$, where $x_n$ is in the direction of the vacuum. Then it follows that, at the boundary,

$$\phi_1 = -2D \frac{\partial \phi_1}{\partial x_n} \qquad (3.17)$$

One way to implement this numerically is to use a linear extrapolation and set $\phi_1$ to be zero at a displaced, virtual boundary that is a distance $1/2D$ into the vacuum from the actual boundary. This displacement, $1/2D$, is known as the *linear extrapolation length*.

### 3.6.2 One-Speed and Two-Speed Approximations

As described earlier, the crudest approach to the energy discretization is to assume that all the neutrons have the same energy, a thermal energy in thermal reactors, because the heat produced is mostly dependent on the thermal neutron flux. This basic approach is termed the *one-speed approximation*, and the diffusion theory based on this approximation is *one-speed diffusion theory*. The next level of approximation is to assume two classes of neutrons each with a single neutron energy. This *two-speed model* applied to thermal reactors assumes one class of thermal neutrons and a second class of fast neutrons combined with a model for the slowing down of the fast neutrons to the thermal neutrons.

It is appropriate here to focus first on the simplest approach, namely, the one-speed approximation. With this approximation, scattering between energy levels is no longer an issue, and the fourth (or [D]) term in the neutron continuity Equation 3.9 drops out. The result is the following governing equation for the neutron flux, $\phi(x_i, t)$ (where the independent variable, $E$, is now dropped because all neutrons are assumed to have the same speed):

$$\frac{1}{\bar{u}} \frac{\partial \phi}{\partial t} - \frac{\partial}{\partial x_j} \left( D(x_i) \frac{\partial \phi}{\partial x_j} \right) + \Sigma_a \phi = S(x_i, t) \qquad (3.18)$$

This is called the *one-speed neutron diffusion equation*, and its solution is known as *one-speed diffusion theory*.

Before moving to examine this theory in some detail, note that the ratio, $D/\Sigma_a$, is a key parameter that appears in the diffusion equation 3.18. The square root of this ratio has the dimension of length and allows the definition of a quantity, $L$, known as the *neutron diffusion length*:

$$L = \left[\frac{D}{\Sigma_a}\right]^{\frac{1}{2}} \tag{3.19}$$

A neutron moving within an absorbing and scattering medium will exhibit classical random walk, and, by Rayleigh's scattering theory, a single neutron will therefore typically travel a distance of $6^{\frac{1}{2}}L$ before it is absorbed. Typical values for $L$ at normal temperatures are of the order of 60 cm. Note that this is not small compared with the dimensions of a reactor core, and therefore diffusion theory can only provide a crude (but nevertheless useful) approximation for reactor neutronics.

### 3.6.3 Steady State One-Speed Diffusion Theory

The most elementary application of diffusion theory is to the steady state operation of a reactor in which the neutron flux is neither increasing or decreasing in time. Then, with the time-derivative term set equal to zero, the one-speed diffusion equation 3.18 becomes

$$-D \nabla^2 \phi = S - \Sigma_a\phi \tag{3.20}$$

assuming that the diffusion coefficient, $D$, is uniform throughout the reactor. Here the left-hand side is the flux of neutrons out of the control volume per unit volume. Thus, in steady state, this must be equal to the right-hand side, the excess of the rate of neutron production over the rate of neutron absorption per unit volume. This excess is a basic property of the fuel and other material properties of the reactor, in other words, a *material property* as defined in Section 2.10. Furthermore, by definition, this excess must be proportional to $(k_\infty - 1)$ (not $(k - 1)$ because the loss to the surroundings is represented by the left-hand side of Equation 3.20). Consequently, it follows that the appropriate relation for the source term is

$$S = k_\infty\Sigma_a\phi \tag{3.21}$$

so that, using the relation 3.19, the one-speed diffusion equation, Equation 3.20, can be written as

$$\nabla^2 \phi + \frac{(k_\infty - 1)}{L^2}\phi = 0 \tag{3.22}$$

The material parameter $(k_\infty - 1)/L^2$ is represented by $B_m^2$ and, as indicated in Section 2.10, is called the *material buckling*:

$$B_m^2 = \frac{(k_\infty - 1)\Sigma_a}{D} = \frac{(k_\infty - 1)}{L^2} \tag{3.23}$$

where $(B_m)^{-1}$ has the dimensions of length. Thus the diffusion equation 3.22 that applies to the steady state operation of the reactor is written as

$$\nabla^2 \phi + B_m^2\phi = 0 \tag{3.24}$$

Equation 3.24 (or 3.22) is Helmholtz' equation. It has convenient solutions by separation of variables in all the simple coordinate systems. Later, detailed eigensolutions to Equation 3.24 will be examined for various reactor geometries. These solutions demonstrate that, in any particular reactor geometry, solutions that satisfy the necessary boundary conditions only exist for specific values (eigenvalues) of the parameter $B_m^2$. These specific values are called the *geometric buckling* and are represented by $B_g^2$; as described in Section 2.10, the values of $B_g^2$ are only functions of the geometry of the reactor and not of its neutronic parameters. It follows that steady state critical solutions only exist when

$$B_m^2 = B_g^2 \qquad (3.25)$$

and this defines the conditions for steady state criticality in the reactor. Moreover, it follows that supercritical and subcritical conditions will be defined by the inequalities

$$\text{Subcritical condition:} \quad B_m^2 < B_g^2 \qquad (3.26)$$

$$\text{Supercritical condition:} \quad B_m^2 > B_g^2 \qquad (3.27)$$

because, in the former case, the production of neutrons is inadequate to maintain criticality and, in the latter, it is in excess of that required.

As a footnote, the multiplication factor, $k$, in the finite reactor can be related to the geometric buckling as follows. From Equation 2.1, $k$ may be evaluated as

$$k = \frac{\text{Rate of neutron production}}{\text{Sum of rates of neutron absorption and escape}} \qquad (3.28)$$

and, in the diffusion equation solution, the rate of escape to the surroundings is represented by $-D \nabla^2 \phi$ and therefore by $D B_g^2 \phi$. The corresponding rate of production is given by $D k_\infty \phi / L^2$ and the rate of neutron absorption by $D\phi / L^2$. Substituting these expressions into Equation 3.28, it is observed that in steady state operation,

$$k = \frac{D k_\infty \phi / L^2}{(D\phi / L^2) + D B_g^2 \phi} = \frac{k_\infty}{(1 + B_g^2 L^2)} \qquad (3.29)$$

### 3.6.4 Two-Speed Diffusion Theory

The next level of approximation is to assume that there are two speeds of neutrons, namely, one group of fast neutrons that are all traveling at the same speed and a second group of thermal neutrons also all traveling with the same speed. These two neutron fluxes will be denoted by $\phi_F$ and $\phi_T$, respectively, and the slowing down from the fast to the thermal neutron group will be modeled by defining a macroscopic cross section for slowing denoted by $\Sigma_{FT}$. Focusing first on the diffusion equation for the thermal neutron flux, $\phi_T$, the source term in Equation 3.20 represents the rate of supply of thermal neutrons due to the slowing down of fast neutrons and will therefore be given by $P_F \Sigma_{FT} \phi_F$, and the first of the two coupled differential equations that constitute the two-speed diffusion model becomes

$$\nabla^2 \phi_T - \frac{\phi_T}{L_T^2} = -\frac{P_F \Sigma_{FT}}{D_T} \phi_F \qquad (3.30)$$

where $L_T$ is the neutron diffusion length for the thermal neutrons.

Turning to the fast neutrons, the absorption of fast neutrons will be neglected in comparison with the slowing down. Then $\Sigma_{FT}$ is analogous to $\Sigma_a$ for the thermal neutrons. Hence a neutron diffusion length for the fast neutrons can be defined as $L_F^2 = D_F / \Sigma_{FT}$. It remains to establish the source term for the fast neutrons, the rate at which fast neutrons are produced by fission. Beginning with the expression 3.21 for $S$ from the one-speed model, it is reasonable to argue that the appropriate $\phi$ in this two-speed model is $\phi_T / P_F$, or the flux of thermal neutrons causing fission in the absence of resonant absorption. Thus the source term in the fast neutron continuity equation will be $k_\infty \Sigma_a \phi_T / P_F$ and the second of the two coupled differential equations, namely, that for the fast neutrons, becomes

$$\nabla^2 \phi_F - \frac{\phi_F}{L_F^2} = -\frac{k_\infty \Sigma_a}{D_F P_F} \phi_T \qquad (3.31)$$

Because $\Sigma_a = D_T / L_T^2$ and $\Sigma_{FT} = D_F / L_F^2$, the two equations 3.31 and 3.30 may be written as

$$\nabla^2 \phi_F - \frac{\phi_F}{L_F^2} = -\frac{D_T}{D_F} \frac{k_\infty}{P_F L_T^2} \phi_T \qquad (3.32)$$

$$\nabla^2 \phi_T - \frac{\phi_T}{L_T^2} = -\frac{D_F}{D_T} \frac{P_F}{L_F^2} \phi_F \qquad (3.33)$$

The solution of these coupled differential equations is simpler than might first appear, for it transpires that the solutions for $\phi_F$ and $\phi_T$ take the same functional form as those of the one-speed equation 3.24 provided the constant $B_g$ is appropriately chosen. This tip-off suggests a solution of the form

$$\nabla^2 \phi_F = -B_g^2 \phi_F \qquad \nabla^2 \phi_T = -B_g^2 \phi_T \qquad (3.34)$$

Substituting into Equations 3.32 and 3.33, it transpires that $B_g^2$ must satisfy

$$(1 + B_g^2 L_T^2)(1 + B_g^2 L_F^2) = k_\infty \qquad (3.35)$$

Consequently, solutions to the two-speed diffusion equations are of the form given in Equations 3.34, where $B_g^2$ must satisfy the quadratic relation 3.35. It follows from Equation 3.35 that there are two possible values for $B_g^2$. In the cases of interest, $k_\infty > 1$, and therefore one of the values of $B_g^2$ is positive and the other is negative. In most circumstances (though not all), the component of the solution arising from the negative root can be neglected or eliminated, leaving only the component resulting from the positive root. Moreover, in the common circumstance in which $k_\infty$ is just slightly greater than unity, the positive root is given approximately by

$$B_g^2 \approx k_\infty / (L_T^2 + L_F^2) \qquad (3.36)$$

Thus both the fission and thermal neutrons are governed by the same diffusion equation as in the one-speed diffusion theory and with a geometric buckling that is a minor modification of that used in the earlier theory. It follows that the one-speed solutions detailed in Sections 3.7.1–3.7.4 can be readily adapted to two-speed solutions.

### 3.6.5 Nonisotropic Neutron Flux Treatments

Before proceeding with derivations from the one- and two-speed diffusion theories, it is appropriate to pause and comment on the many approximations that were made in developing these models and to outline the more accurate efforts that are required for detailed reactor analysis and design. In reviewing the extensive assumptions that were made in the preceding sections, it is surprising that the simple diffusion theories work at all; indeed, to the extent that they do, that success is largely a result of judicious choice of the averaging used to arrive at the effective cross sections.

One set of assumptions was that the angular neutron flux was isotropic (or nearly so). Several approaches have been developed to model anisotropy, the deviations from this isotropy. One is to select a number, $N$, of angular directions and to represent the neutron flux as the sum of fluxes in each of these discrete directions. This leads to a set of neutron transport equations, one for each of the discrete directions. These are known as the $S_N$ equations. A preferred alternative is to represent the anisotropy using a finite series of $N$ spherical harmonic functions and to develop neutron transport equations for each of the terms in this series. These are known as the $P_N$ equations, the most commonly used having just one nonisotropic term ($P_1$ equations). However, in many circumstances in most large reactors, the assumption of an isotropic neutron flux is reasonably valid, except perhaps in the neighborhood of nonisotropic material (that is, for example, highly absorbing) or at a boundary that results in a highly nonisotropic neutron flux. Such regions or boundaries can then be given special treatment using one of the previously mentioned approaches.

### 3.6.6 Multigroup Diffusion Theories and Calculations

A second and more important set of assumptions was the very limited discretization of the energy spectrum. Perhaps the most glaring deficiency of the one-speed diffusion theory is the assumption that all the neutrons have the same speed or energy. Consequently, the most obvious improvement would be to allow a variety of neutron energies and to incorporate a model for the transfer of neutrons from one energy level to another. Such models are termed *multigroup diffusion models*, and the simplest among these is the *two-speed diffusion model* in which the neutron population consists of one population of fast neutrons and another of thermal neutrons. This is particularly useful in a LWR in which the moderator helps maintain a balance between the two groups. More sophisticated models with many more energy levels are needed to accommodate the complexities of the neutron energy spectra described in Section 2.3.2. As illustrated in Figure 2.3, the variations in the cross sections (and source terms) over the neutron energy range can be very complicated. It is therefore impractical to devise a neutron transport treatment that accurately incorporates all of these variations. In detailed, practical calculations, a compromise is necessary, and the energy spectrum is often divided into 20 or 30 energy levels (or *groups*); in other words, it is divided much more finely than in the one- or two-speed model. However, because 20 or 30 groups still cannot adequately cover the variations of the cross sections with energy level, it is necessary to devise averaging methods within each range or group to obtain *effective* cross sections and source terms that adequately represent the neutron behavior within that group. It is evident that these methods and calculations are only as good as the accuracy of the source terms and cross sections assumed. Therefore careful analysis and modeling of the scattering process is critical, as is accurate representation and averaging of the cross sections within each energy level. Diffusion equations have been developed for each of these energy groups, the approach being called *multigroup diffusion models*. Sophisticated numerical schemes have been developed for the solution of all these coupled differential equations (see, e.g., Glasstone and Sesonske 1981; Duderstadt and Hamilton 1976), and modern reactor designs rely on these detailed calculations, which are beyond the scope of this monograph.

### 3.6.7 Lattice Cell Calculations

A third set of assumptions involves the averaging over the various materials that make up a reactor core. The fuel rods, control rods, moderator, coolant channels, and so on, in a reactor are usually arranged in *lattice cells* that are repeated across the cross section of the core (see Section 4.3.4 and Figure 4.10). Thus there are several structural or material scales (dimensions), a small common scale being the diameter of the fuel rods. Another, larger scale would be the dimension of the lattice. In the preceding sections, it was assumed that the core was effectively homogeneous; this is the case when the material inhomogeneity dimension is small compared with the typical mean free path of the neutrons. In a light water reactor (LWR), the typical mean free path is of the order of centimeters and therefore comparable with the diameter of a fuel rod. In contrast, a fast breeder reactor has typical mean free paths of the order of tens of centimeters but similar fuel rod dimensions, and so the inhomogeneity is less important in fast reactor calculations.

When, as in a LWR, the inhomogeneity is important, there will be a significant difference between the neutron flux within the fuel rods and that in the moderator or coolant. Practical reactor analysis and design require detailed calculation of these differences, and this is effected using numerical codes called *heterogeneous lattice cell calculations*. Approximate diffusion theory methods used to evaluate these inhomogeneity effects in real reactors are briefly discussed in Section 3.8.

### 3.7 Simple Solutions to the Diffusion Equation

### 3.7.1 Spherical and Cylindrical Reactors

Notwithstanding the limitations of the one-speed diffusion theory, it is appropriate to pursue further reactor analyses because they yield qualitatively useful results and concepts. As previously mentioned, the Helmholtz diffusion equation 3.24 permits solutions by separation of variables in many simple coordinate systems. Perhaps the most useful are the solutions in cylindrical coordinates because this closely approximates the geometry of most reactor cores.

However, the solutions in spherical coordinates are also instructive, and it is useful to begin with these. It is readily seen that, in a spherically symmetric core (radial coordinate, $r$), the solution to Equation 3.24 takes the form

$$\phi = C_1 \frac{\sin B_g r}{r} + C_2 \frac{\cos B_g r}{r} \tag{3.37}$$

where $C_1$ and $C_2$ are constants to be determined. For $\phi$ to be finite in the center, $C_2$ must be zero. The boundary condition at the surface, $r = R$, of this spherical reactor follows from the assumption that it is surrounded by a vacuum. Consequently, the appropriate boundary condition is given by Equation 3.17, or more conveniently $\phi = 0$, at the extrapolated boundary at $r = R_E = R + 1/2D$. Thus

$$\sin B_g R_E = 0 \qquad \text{or} \qquad B_g R_E = n\pi \tag{3.38}$$

where $n$ is an integer. Because $B_g$ and $n$ are positive and $\phi$ cannot be negative anywhere within the core, the only acceptable, nontrivial value for $n$ is unity, and therefore

$$R_E = \pi / B_g \qquad \text{and thus} \qquad R = \pi / B_g - 1/2D \tag{3.39}$$

Therefore, $R = R_C = \pi / B_m - 1/2D$ is the critical size of a spherical reactor, that is to say, the only size for which a steady neutron flux state is possible for the given value of the material buckling, $B_m$. It is readily seen from Equation 3.9 that $\partial \phi / \partial t$ will be positive if $R > R_C$ and that the neutron flux will then grow exponentially with time. Conversely, when $R < R_C$, the neutron flux will decay exponentially with time.

In summary, the neutron flux solution for the steady state operation of a spherically symmetric reactor is

$$\phi = C_1 \frac{\sin B_g r}{r} \qquad \text{for} \qquad 0 < r < R_C \tag{3.40}$$

Note that the neutron flux is largest in the center and declines near the boundary owing to the increased leakage. Also note that though the functional form of the neutron flux variation has been determined, the magnitude of the neutron flux as defined by $C_1$ remains undetermined because the governing equation and boundary conditions are all homogeneous in $\phi$.

Most common reactors are cylindrical, and so, as a second example, it is useful to construct the solution for a cylinder of radius, $R$, and axial length, $H$, using cylindrical coordinates, $(r, \theta, z)$, with the origin at the mid-length of the core. It is assumed that the reactor is homogeneous so that there are no gradients in the $\theta$ direction and both the sides and ends see vacuum conditions. Again, it is convenient to apply the condition $\phi = 0$ on extrapolated boundary surfaces at $r = R_E = R + 1/2D$ and

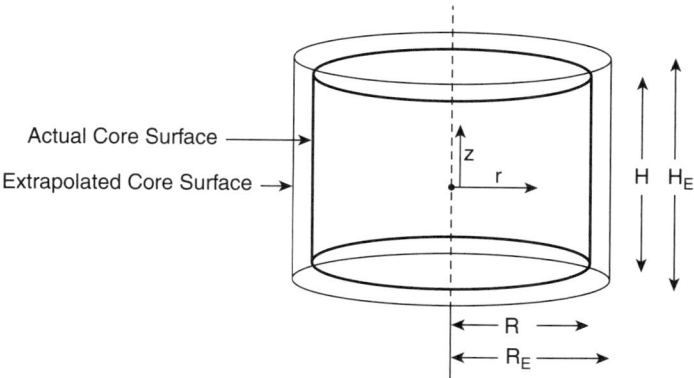

Figure 3.1. Sketch of a simple cylindrical reactor.

at $z = \pm H_E/2 = \pm(H/2 + 1/2D)$, as depicted in Figure 3.1. Obtaining solutions to Equation 3.24 by separation of variables and eliminating possible solutions that are singular on the axis, it is readily seen that the neutron flux has the form

$$\phi = C_1 \cos\left(\frac{\pi z}{H_E}\right) J_0\left(\frac{2.405r}{R_E}\right) \tag{3.41}$$

where, as before, $C_1$ is an undetermined constant and $J_0()$ is the zero-order Bessel function of the first kind (2.405 is the argument that gives the first zero of this function). As in the spherical case, the higher-order functions are rejected because they would imply negative neutron fluxes within the cylindrical reactor. Substituting this solution into the governing equation 3.24 yields the expression that determines the critical size of this cylindrical reactor, namely,

$$\left(\frac{\pi}{H_E}\right)^2 + \left(\frac{2.405}{R_E}\right)^2 = B_g^2 \tag{3.42}$$

If $H_E$ and $R_E$ are such that the left-hand side is greater than the material buckling, $B_m^2$, then the reactor is supercritical and the neutron flux will grow exponentially with time; if the left-hand side is less than $B_m^2$, the flux will decay exponentially. In the critical reactor ($B_g^2 = B_m^2$), the neutron flux is greatest in the center and decays toward the outer radii or the ends because the leakage is greatest near the boundaries.

These two examples assumed homogeneous reactors surrounded by vacuum conditions. These simple solutions can be modified in a number of ways to incorporate common, practical variations. Often the reactor core is surrounded, not by a vacuum, but by a *blanket* of moderator that causes some of the leaking neutrons to be scattered back into the core. Such a blanket is called a reflector; examples of diffusion theory solutions that incorporate the effect of a reflector are explored in the next section. Another practical modification is to consider two core regions rather than one to model that region into which control rods have been inserted. Section 3.7.4 includes an example of such a two-region solution.

### 3.7.2 Effect of a Reflector on a Spherical Reactor

In the examples of the last section, it was assumed that all neutrons leaking out were lost. In practice, reactor cores are usually surrounded by a reflector that scatters some of the leaking neutrons back into the core. In this section, two examples of diffusion theory solutions with reflectors are detailed.

Perhaps the simplest example is the spherically symmetric reactor of the preceding section, now surrounded by a reflector of inner radius $R$ and outer radius $R_R$. Because there is no source of neutrons in the reflector, the diffusion equation that governs the neutron flux in the reflector (denoted by $\phi_R$) is then

$$\nabla^2 \phi_R - \frac{1}{L_R^2}\phi_R = 0 \qquad (3.43)$$

where $L_R$ is the diffusion length in the reflector. The boundary conditions that must be satisfied are as follows. At the interface between the core and the reflector, both the neutron flux and the net radial neutron current (see Section 3.2) must match so that

$$(\phi)_{r=R} = (\phi_R)_{r=R} \qquad D\left(\frac{\partial \phi}{\partial r}\right)_{r=R} = D_R\left(\frac{\partial \phi_R}{\partial r}\right)_{r=R} \qquad (3.44)$$

where $D$ and $D_R$ are the diffusion coefficients in the core and in the reflector. At the outer boundary of the reflector, the vacuum condition requires that $\phi_R = 0$ at $r = R_R + 1/2D_R = R_{RE}$.

As in the preceding section, the appropriate solution for the neutron flux in the core is

$$\phi = \frac{C}{r}\sin B_g r \qquad (3.45)$$

where $C$ is an undetermined constant. Moreover, the appropriate solution to Equation 3.43 in the reflector is

$$\phi_R = \frac{C_R}{r}\sinh\left(\frac{r^* - r}{L_R}\right) \qquad (3.46)$$

where $C_R$ and $r^*$ are constants as yet undetermined. Applying the preceding boundary conditions, it follows that

$$r^* = R_{RE} \qquad C\sin B_g R = C_R \sinh\left(\frac{R_{RE} - R}{L_R}\right) \qquad (3.47)$$

and that

$$DC(\sin B_g R - B_g R \cos B_g R) \qquad (3.48)$$

$$= D_R C_R \left(\frac{R}{L_R}\cosh\left(\frac{R_{RE} - R}{L_R}\right) + \sinh\left(\frac{R_{RE} - R}{L_R}\right)\right)$$

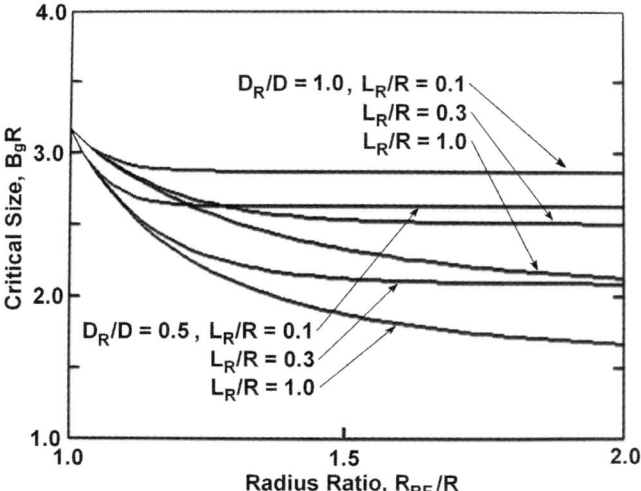

Figure 3.2. The nondimensional critical size or geometric buckling, $B_g R$, for a spherical reactor with a reflector as a function of the radius ratio, $R_{RE}/R$, for various values of $L_R/R$ and $D_R/D$.

Eliminating the ratio $C/C_R$ from the last two relations yields

$$D(1 - B_g R \cot B_g R) = D_R \left(1 + \frac{R}{L_R} \coth\left(\frac{R_{RE} - R}{L_R}\right)\right) \qquad (3.49)$$

Given all the material constants involved, this relation can be solved numerically to determine the critical size (or critical geometric buckling) of such a spherical reactor.

Sample results are shown in Figure 3.2, which presents the nondimensional critical size or geometric buckling, $B_g R$, as a function of the radius ratio, $R_{RE}/R$, for various values of $L_R/R$ and $D_R/D$. The change in the shape of the neutron flux as the size of the reflector is increased, is shown in Figure 3.3; note that the uniformity of the neutron flux within the core can be somewhat improved by the presence of the reflector.

Figure 3.3. The shape of the neutron flux distribution in a spherical reactor surrounded by a reflector, $\phi$ (normalized by the maximum neutron flux, $\phi_M$), for various radius ratios, $R_{RE}/R$, as shown and for $D_R/D = 1$.

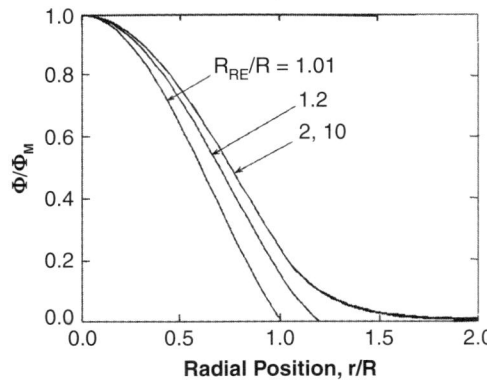

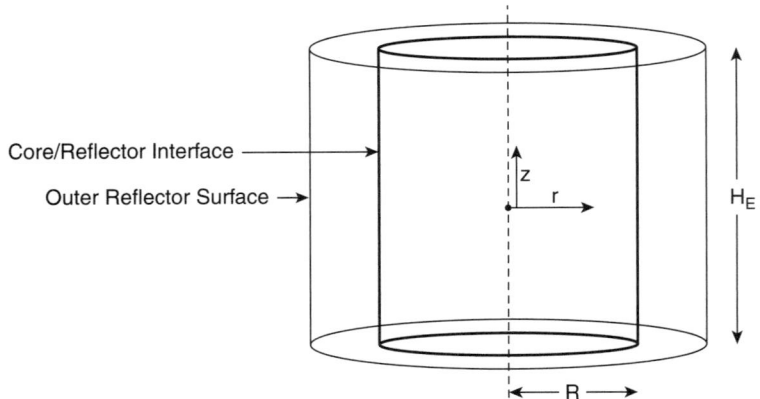

Figure 3.4. Cylindrical reactor with reflector.

### 3.7.3  Effect of a Reflector on a Cylindrical Reactor

As a second example of the effect of a reflector, consider the cylindrical reactor of Section 3.7.1 surrounded at larger radii by a reflector as shown in Figure 3.4 (for simplicity it is assumed that vacuum conditions pertain at both ends of the core and the reflector). Then, as in Section 3.7.1, the appropriate, nonsingular solution to Equation 3.24 for the neutron flux in the core is

$$\phi = C \cos\left(\frac{\pi z}{H_E}\right) J_0(\xi_1 r) \qquad (3.50)$$

where $H_E = H + 1/D$ as before and $C$ and $\xi_1$ are constants as yet undetermined. Turning now to the solution for Equation 3.43 in the cylindrical annulus occupied by the reflector, it is assumed, for simplicity, that this extends all the way from $r = R$ to $r \to \infty$ and that the reflector has the same height $H_E$ as the core. Then, omitting terms that are singular as $r \to \infty$, the appropriate solution to Equation 3.43 in the reflector is

$$\phi_R = C_R \cos\left(\frac{\pi z}{H_E}\right) K_0(\xi_2 r) \qquad (3.51)$$

where $\xi_2$ is to be determined and $K_0$ is the modified Bessel function. Applying the boundary conditions at the core-reflector interface, $r = R$ (Equations 3.44), yields the relations

$$CJ_0(\xi_1 R) = C_R K_0(\xi_2 R) \qquad \xi_1 D C J_1(\xi_1 R) = \xi_2 D_R C_R K_1(\xi_2 R) \qquad (3.52)$$

and, upon elimination of $C_R/C$, these yield

$$D\xi_1 J_1(\xi_1 R) K_0(\xi_2 R) = D_R \xi_2 K_1(\xi_2 R) J_0(\xi_1 R) \qquad (3.53)$$

and, in a manner analogous to Equation 3.49, this equation must be solved numerically to determine $R$, the critical size of such a cylindrical reactor. The corresponding solutions for a reflector with a finite outer radius or with a reflector at the ends, though algebraically more complicated, are conceptually similar.

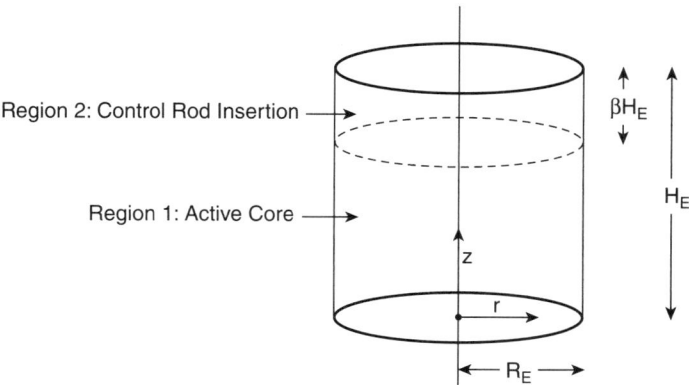

Figure 3.5. Cylindrical reactor with partial control rod insertion.

### 3.7.4 Effect of Control Rod Insertion

A second example of a practical modification of the diffusion theory solutions is to consider a core into which control rods have been partially inserted so that, as sketched in Figure 3.5, the reactor core consists of two regions with different levels of neutron absorption. The fractional insertion will be denoted by $\beta$. Assuming that the control rod absorption is sufficiently large so that the conditions in the controlled region are subcritical, the equations governing the neutron flux in the two regions are

$$\nabla^2 \phi_1 + B_g^2 \phi_1 = 0 \quad \text{in} \quad 0 \le z \le (1 - \beta)H_E \tag{3.54}$$

$$\nabla^2 \phi_2 - \frac{\phi_2}{L_2^2} = 0 \quad \text{in} \quad (1 - \beta)H_E \le z \le H_E \tag{3.55}$$

where subscripts 1 and 2 refer to the two regions indicated in Figure 3.5, $L_2$ is the neutron diffusion length in region 2, and, for convenience, the origin of $z$ has been shifted to the bottom of the core. The boundary conditions on the cylindrical surface $r = R_E$ are $\phi_1 = \phi_2 = 0$ (as in Section 3.7.1), and on the radial planes they are

$$\phi_1 = 0 \quad \text{on} \quad z = 0 \qquad \phi_2 = 0 \quad \text{on} \quad z = H_E \tag{3.56}$$

$$\phi_1 = \phi_2 \quad \text{and} \quad \frac{\partial \phi_1}{\partial z} = \frac{\partial \phi_2}{\partial z} \quad \text{on} \quad z = (1 - \beta)H_E \tag{3.57}$$

where, for simplicity, it has been assumed that the neutron diffusivities are the same in both regions. By separation of variables, the appropriate solutions to Equations 3.54 and 3.55 are

$$\phi_1 = [C_1 \sin \xi_1 z + C_2 \cos \xi_1 z] J_0(2.405r/R_E) \tag{3.58}$$

$$\phi_2 = [C_3 e^{\xi_2 z} + C_4 e^{-\xi_2 z}] J_0(2.405r/R_E) \tag{3.59}$$

where $C_1, C_2, C_3, C_4, \xi_1,$ and $\xi_2$ are constants as yet undetermined and the boundary conditions at $r = R_E$ have already been applied. The governing equations 3.54

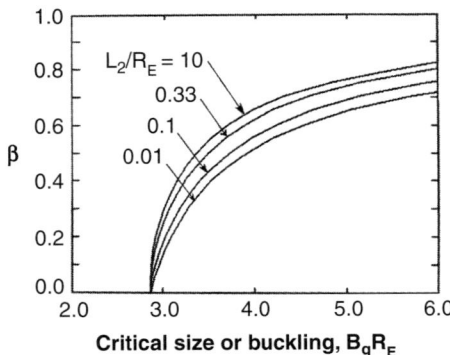

Figure 3.6. The critical nondimensional size or geometric buckling, $B_g R_E$, as a function of the fractional control rod insertion, $\beta$, for a cylindrical reactor with $H_E/R_E = 2.0$ and several values of $L_2/R_E$, as indicated.

and 3.55 require that

$$\xi_1^2 = B_g^2 - (2.405/R_E)^2 \qquad \xi_2^2 = (1/L_2)^2 + (2.405/R_E)^2 \tag{3.60}$$

The boundary conditions 3.56 require that

$$C_2 = 0 \qquad C_4 = -C_3 e^{2\xi_2 H_E} \tag{3.61}$$

and using these with the boundary conditions 3.57 yields

$$C_1 \sin\left\{\xi_1(1-\beta)H_E\right\} = -C_3 e^{\xi_2 H_E}\left[e^{\xi_2 \beta H_E} - e^{-\xi_2 \beta H_E}\right] \tag{3.62}$$

$$\xi_1 C_1 \cos\left\{\xi_1(1-\beta)H_E\right\} = \xi_2 C_3 e^{\xi_2 H_E}\left[e^{\xi_2 \beta H_E} + e^{-\xi_2 \beta H_E}\right] \tag{3.63}$$

Eliminating the ratio $C_1/C_3$ from these last two expressions yields

$$\xi_2 \tan\left\{\xi_1(1-\beta)H_E\right\} + \xi_1 \tanh\left\{\xi_2 \beta H_E\right\} = 0 \tag{3.64}$$

Because $\xi_1$ and $\xi_2$ are given by Equations 3.60, this constitutes an expression for the critical size of the reactor, $R_E$ (or $R$), given the aspect ratio $H_E/R_E$ as well as $B_g$, $L_2$, and $\beta$. Equivalently, it can be seen as the value of $\beta$ needed to generate a critical reactor given $R_E$, $H_E$, $B_g$, and $L_2$.

As a nondimensional example, Figure 3.6 presents critical values for the fractional insertion, $\beta$, as a function of the quantity $B_g R_E$ (which can be thought of as a nondimensional size or nondimensional geometric buckling) for a typical aspect ratio, $H_E/R_E$, of 2.0 and several values of $L_2/R_E$. Naturally, the critical size increases with the insertion, $\beta$; equivalently, the insertion, $\beta$, for a critical reactor will increase with the *size* given by $B_g R_E$. Note that the results are not very sensitive to the value of $L_2/R_E$.

The way in which the neutron flux distribution changes as the control rods are inserted will become important when the temperature distribution is analyzed in later chapters. Evaluating the neutron flux in the preceding solution and normalizing each distribution in the $z$ direction by the maximum value of $\phi$ occurring within it

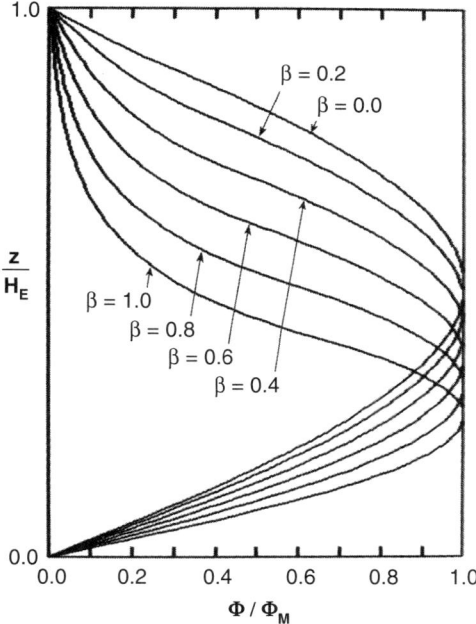

Figure 3.7. The change in the shape of the axial distribution of the neutron flux, $\phi$ (normalized by the maximum neutron flux, $\phi_M$), with fractional control rod insertion, $\beta$, for the case of $H_E/R_E = 2.0$ and $L_2/R_E = 0.36$.

(denoted by $\phi_M$), the distribution becomes

$$\phi/\phi_M = \sin\{\xi_1 z\} \quad \text{for} \quad 0 \le z \le (1-\beta)H_E$$

$$= \frac{\sin\{\xi_1(1-\beta)H_E\}}{\{e^{\xi_2 \beta H_E} - e^{-\xi_2 \beta H_E}\}} \{e^{\xi_2(H_E-z)} - e^{-\xi_2(H_E-z)}\}$$

$$\text{for} \quad (1-\beta)H_E \le z \le H_E \tag{3.65}$$

Typical examples of these neutron flux distributions are shown in Figure 3.7; as the fractional insertion, $\beta$, increases, note how the neutron flux in the region of insertion decreases and the distribution becomes skewed toward the lower part of the core.

## 3.8 Steady State Lattice Calculations

### 3.8.1 Introduction

The preceding sections addressed solutions for the distribution of the neutron flux in reactors that were assumed to be homogeneous. As previously described in Section 3.6.1, the assumption of homogeneity was based on the fact that the typical mean free path of the neutrons is normally large compared with the small-scale structure within the reactor (e.g., the fuel rod diameter). Conversely, the neutron mean free path is somewhat smaller than the overall reactor geometry, and this provides some qualitative validity for variations in the neutron flux within the reactor that are derived from analytical methods such as those based on the diffusion equation (or any other more detailed methodology).

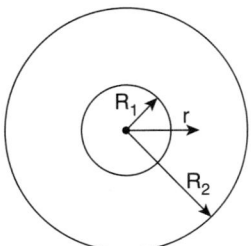

Figure 3.8. Generic circular lattice cell.

As described in Section 3.6.7, the typical mean free path in a LWR is of the order of centimeters and therefore comparable with the diameter of a fuel rod. Consequently, the variation of the neutron flux within and around the fuel rod of a LWR may be substantial and is therefore important to take into account in the design of those components. In contrast, a fast breeder reactor has typical mean free paths of the order of tens of centimeters. With fuel rod dimensions similar to a LWR, it follows that the inhomogeneity is less important in a fast reactor. In either case, practical reactor analysis and design require detailed calculation of the variations in the neutron flux at these smaller scales, and this can be effected using numerical codes called *heterogeneous lattice cell calculations*.

Thus it is appropriate to consider analytical methods that might be used to determine the variations in the neutron flux associated with the finer structure within a reactor core, for example, the variations around a fuel rod or a control rod. In this endeavor, it is convenient to take advantage of the fact that much of this finer structure occurs in lattices or *units* that are repeated over the cross section of the reactor (or at least parts of that cross section). For example, each of the fuel rods is surrounded by coolant channels and other fuel rods in patterns that are exemplified in Figure 4.10. Thus a fuel rod plus an appropriately allocated fraction of the surrounding coolant constitute a *unit*, and those units are repeated across the reactor cross section. Moreover, because the neutron mean free path is comparable to or larger than the dimensions of this unit, it may be adequate to adjust the geometry of the unit to facilitate the mathematical solution of the neutron flux. Thus, as shown in Figure 3.8, the geometry of a fuel rod unit might be modeled by a central cylinder of fuel pellets of radius, $R_1$, surrounded by a cylinder of neutronically passive and moderating material, $R_1 < r < R_2$, where the ratio of the areas of the two regions is the same as the ratio of the cross-sectional area of fuel pellet to the cross-sectional area of allocated nonpellet material within the reactor.

Before outlining some typical examples of the calculation of the neutron flux variations within a lattice cell, it is necessary to consider the nature of the boundary conditions that might be applied at the interfaces and boundaries of a cell such as that of Figure 3.8. As described in Section 3.6.1, at an interface such as $r = R_1$, not only should the neutron fluxes in the two regions be the same but the one-way fluxes must also be the same. In the context of diffusion theory, these imply that at the interface, the conditions should be as given in Equation 3.16. With the geometry of Figure 3.8, these become

$$\phi_1 = \phi_2 \quad \text{and} \quad D_1 \frac{\partial \phi_1}{\partial r} = D_2 \frac{\partial \phi_2}{\partial r} \tag{3.66}$$

Now consider the conditions on the outer boundary of the lattice cell ($r = R_2$ in Figure 3.8). If the reactor is in a steady critical state, each of the unit cells should be operating similarly with little or no net neutron exchange between them, and therefore the condition on the outer boundary should be

$$\frac{\partial \phi_2}{\partial r} = 0 \quad \text{on} \quad r = R_2 \tag{3.67}$$

or the equivalent in more complex neutron flux models.

In the sections that follow, diffusion theory solutions are used to explore some of the features of these lattice cell models.

### 3.8.2 Fuel Rod Lattice Cell

The diffusion theory solution for the single fuel rod lattice cell requires the solution of the following forms of the diffusion equation 3.22 for the neutron fluxes $\phi_1$ and $\phi_2$ in the two regions of the cell sketched in Figure 3.8:

$$\nabla^2 \phi_1 + B_g^2 \phi_1 = 0 \quad \text{in} \quad r \leq R_1 \tag{3.68}$$

$$\nabla^2 \phi_2 - \frac{\phi_2}{L_2^2} = 0 \quad \text{in} \quad R_1 \leq r \leq R_2 \tag{3.69}$$

subject to the boundary conditions 3.66 and 3.67 and neglecting any gradients in the direction normal to Figure 3.8 sketch (the $z$ direction). The appropriate general solutions are

$$\phi_1 = C_1 J_0(B_g r) + C_2 Y_0(B_g r) \tag{3.70}$$

$$\phi_2 = C_3 I_0(r/L_2) + C_4 K_0(r/L_2) \tag{3.71}$$

where $J_0()$ and $Y_0()$ are Bessel functions of the first and second kind, $I_0()$ and $K_0()$ are modified Bessel functions of the first and second kind, and $C_1, C_2, C_3$, and $C_4$ are constants yet to be determined. Because $\phi_1$ must be finite at $r = 0$, $C_2$ must be zero. If, for the convenience of this example, the diffusivities are assumed to be the same in both regions ($D_1 = D_2 = D$), then the boundary conditions 3.61 require that

$$C_1 J_0(B_g R_1) = C_3 I_0(R_1/L_2) + C_4 K_0(R_1/L_2) \tag{3.72}$$

$$-C_1 B_g J_1(B_g R_1) = C_3 I_1(R_1/L_2)/L_2 - C_4 K_1(R_1/L_2)/L_2 \tag{3.73}$$

where $J_1()$, $I_1()$, and $K_1()$ denote Bessel functions of the first order. In addition, the outer boundary condition 3.62 requires that

$$-C_3 I_1(R_2/L_2) + C_4 K_1(R_2/L_2) = 0 \tag{3.74}$$

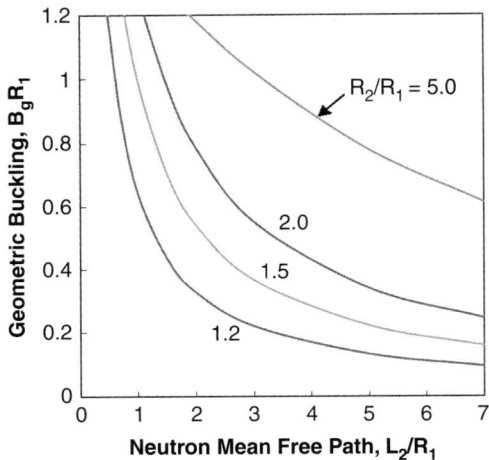

Figure 3.9. Values of the nondimensional geometric buckling for the fuel rod lattice cell as a function of $L_2/R_1$ for four values of $R_2/R_1$, as shown.

and Equations 3.72, 3.73, and 3.74 lead to the eigenvalue equation

$$J_0(B_g R_1)\left[I_1(R_2/L_2)K_1(R_1/L_2) - I_1(R_1/L_2)K_1(R_2/L_2)\right]$$
$$= L_2 B_g J_1(B_g R_1)\left[I_0(R_1/L_2)K_1(R_2/L_2) + I_1(R_2/L_2)K_0(R_1/L_2)\right] \quad (3.75)$$

Given the nondimensional parameters $R_2/R_1$ and $L_2/R_1$, the solution to this equation yields the nondimensional geometric buckling, $B_g R_1$, for this configuration. When the neutron mean free path, $L_2$, is large relative to $R_1$ and $R_2$, the approximate solution to Equation 3.75 is

$$B_g^2 L_2^2 \approx (R_2^2 - R_1^2)/R_1^2 \quad (3.76)$$

More precise solutions for the nondimensional geometric buckling, $B_g^2 R_1^2$, are shown in Figure 3.9 for various values of $L_2/R_1$ and $R_2/R_1$. These lead to different neutron flux profiles, as exemplified by those presented in Figure 3.10. As expected, the flux

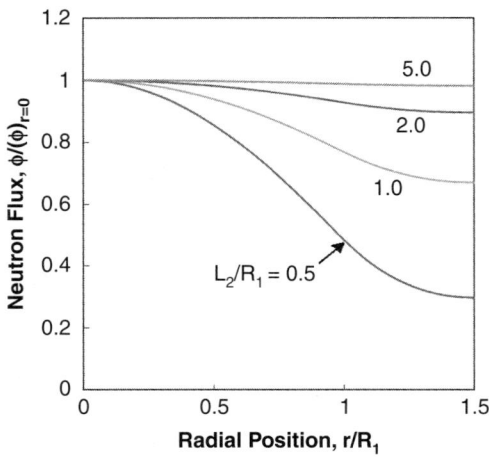

Figure 3.10. Typical neutron flux profiles for the fuel rod lattice cell with $R_2/R_1 = 1.5$ for four values of $L_2/R_1$, as shown.

inside the fuel rod is larger than in the surroundings, but the profile flattens out as the neutron mean free path increases.

### 3.8.3 Control Rod Lattice Cell

The obverse of the fuel rod lattice cell is the control rod lattice cell in which an individual control rod ($0 < r < R_1$) is surrounded by an annulus ($R_1 < r < R_2$) containing a homogeneous mix of fuel rod and coolant, as can also be depicted by Figure 3.8. Then the governing equations for the neutron flux are

$$\nabla^2 \phi_1 - \frac{\phi_1}{L_1^2} = 0 \quad \text{in} \quad r \leq R_1 \tag{3.77}$$

$$\nabla^2 \phi_2 + B_g^2 \phi_2 = 0 \quad \text{in} \quad R_1 \leq r \leq R_2 \tag{3.78}$$

where gradients in the direction normal to the sketch (the $z$ direction) are neglected and $L_1$ is the neutron mean free path in the control rod. The boundary conditions are the same as in the fuel rod lattice cell, and it follows that the appropriate general solutions are

$$\phi_1 = C_1 I_0(r/L_1) + C_2 K_0(r/L_1) \tag{3.79}$$

$$\phi_2 = C_3 J_0(B_g r) + C_4 Y_0(B_g r) \tag{3.80}$$

Because $\phi_1$ must be finite at $r = 0$, $C_2$ must be zero. If, again for convenience, the diffusivities are assumed to be the same in both regions ($D_1 = D_2 = D$), then the boundary conditions 3.61 require that

$$C_1 I_0(R_1/L_1) = C_3 J_0(B_g R_1) + C_4 Y_0(B_g R_1) \tag{3.81}$$

$$C_1 I_1(R_1/L_1)/L_1 = -C_3 B_g J_1(B_g R_1) - C_4 B_g Y_1(B_g R_1) \tag{3.82}$$

In addition, the outer boundary condition 3.62 requires that

$$C_3 J_1(B_g R_2) + C_4 Y_1(B_g R_2) = 0 \tag{3.83}$$

and Equations 3.81, 3.82, and 3.83 lead to the eigenvalue equation

$$I_1(R_1/L_1) \left[ J_0(B_g R_1) Y_1(B_g R_2) - J_1(B_g R_2) Y_0(B_g R_1) \right]$$

$$= B_g L_1 I_0(R_1/L_1) \left[ J_1(B_g R_2) Y_1(B_g R_1) - J_1(B_g R_1) Y_1(B_g R_2) \right] \tag{3.84}$$

Given the nondimensional parameters $R_2/R_1$ and $L_1/R_1$, the solution to this equation yields the nondimensional geometric buckling, $B_g R_1$, for this configuration. When the neutron mean free path, $L_1$, is large relative to $R_1$ and $R_2$, the approximate solution to Equation 3.84 is

$$B_g^2 L_1^2 \approx R_1^2/(R_2^2 - R_1^2) \tag{3.85}$$

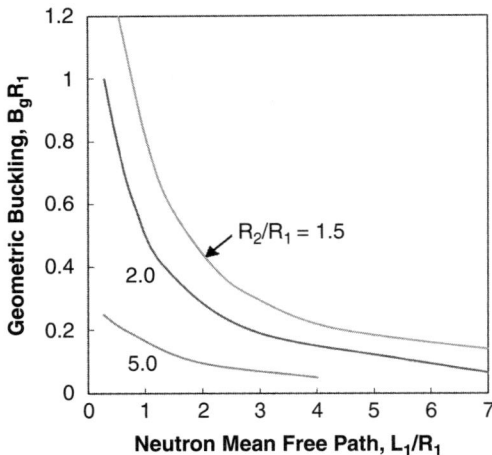

Figure 3.11. Values of the nondimensional geometric buckling for the control rod lattice cell as a function of $L_1/R_1$ for three values of $R_2/R_1$, as shown.

More precise solutions for the nondimensional geometric buckling, $B_g^2 R_1^2$, are shown in Figure 3.11 for various values of $L_1/R_1$ and $R_2/R_1$. These lead to different neutron flux profiles, as exemplified by those presented in Figure 3.12. As expected, the flux inside the control rod ($r < R_1$) is smaller than in the surroundings, but the profile flattens out as the geometric buckling decreases.

### 3.8.4 Other Lattice Scales

The preceding two sections illustrated the use of the lattice cell approach, first on the small scale associated with an individual fuel rod and then on the somewhat larger scale associated with an individual control rod. Finally, it should be noted that many other lattice cell approaches are possible. For example, the square cross section fuel assembly sketched in Figure 3.8 is repeated across a PWR core, and this can be utilized to investigate inhomogeneous effects on that scale. For such square cross section

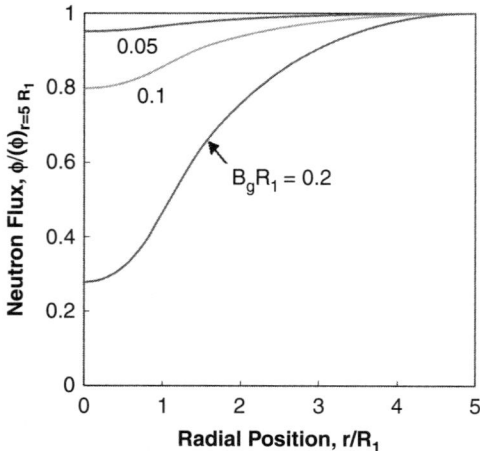

Figure 3.12. Typical neutron flux profiles for the control rod lattice cell with $R_2/R_1 = 5$ for three values of nondimensional geometric buckling, $B_g R_1$, as shown.

lattice cells, the diffusion equation 3.24 has solutions of the form

$$\phi = C \cos \left( B_m x/2^{1/2} \right) \cos \left( B_m y/2^{1/2} \right) \qquad (3.86)$$

where the origin of the $(x, y)$ coordinate system is taken to be the center of the square cross section. Solutions like Equation 3.86 combined with the fact that the diffusion equations permit superposition of solutions allow the construction of a variety of other lattice cell solutions to that equation.

However, it is important to note in closing that these diffusion equation approaches involve many approximations and can only be considered to provide qualitative estimates and guidance. Precise, quantitative assessment of the neutronics of a reactor core are much more complex (see, e.g., Duderstadt and Hamilton 1976) and require much greater computational effort.

## 3.9 Unsteady or Quasi-Steady Neutronics

The preceding sections in this chapter referred to steady state calculations, and therefore some mention of the corresponding time-dependent processes should be made before concluding this brief introduction to reactor core neutronics. Clearly it is important to consider the growth or decay rates for the neutron flux when the reactor becomes supercritical or subcritical. This is needed not only to design control systems for the reactor but also to evaluate scenarios that would follow reactor transients or accidents. Two sets of unsteady perturbations are commonly considered: (1) perturbations caused by changes in the reactor core neutronics, for example, the insertion or withdrawal of control rods, or (2) perturbations caused by changes in the thermohydraulic conditions, such as a change in power level. The former perturbations are governed by what are called the *nuclear reactor kinetics*, whereas the latter are termed the *nuclear reactor dynamics*. The latter therefore involve the response of the entire plant, including the steam generators, and are not further discussed in this text. Instead, attention is confined to the nuclear reactor kinetics.

### 3.9.1 Unsteady One-Speed Diffusion Theory

To exemplify the nature of nuclear reactor kinetics, it is convenient to return to the basic one-speed diffusion equation 3.18. Retaining the unsteady term, $\partial \phi / \partial t$, this becomes

$$\frac{1}{\bar{u}} \frac{\partial \phi}{\partial t} - D \nabla^2 \phi = S - \Sigma_a \phi \qquad (3.87)$$

where it is assumed that the diffusivity is uniform throughout the reactor and does not change with time. Recall that Equation 3.87 is a statement of neutron conservation in some small piece of the core in which the excess of the neutrons produced over the neutrons absorbed (the right-hand side) is balanced by the rate of increase of the neutrons in that piece plus the net flux of neutrons out of that piece of core (the left-hand side). As before, the left-hand side is set equal to $(k_\infty - 1)\Sigma_a \phi$ so that the diffusion equation 3.87 becomes

$$\frac{1}{\bar{u}D}\frac{\partial\phi}{\partial t} - \nabla^2\phi = \frac{(k_\infty - 1)\Sigma_a}{D}\phi \tag{3.88}$$

Fortunately, Equation 3.88 is linear in the neutron flux, $\phi$, and therefore solutions are superposable. Consequently, for simplicity, the focus is on a single basic solution, knowing that more complex solutions may be constructed by superposition. This basic solution for the neutron flux, $\phi(x_i, t)$, takes the form

$$\phi(x_i, t) = C\exp(-\xi t)\phi^*(x_i) \tag{3.89}$$

where $C$ is a constant, $\xi$ is the time constant associated with the transient, and $\phi^*$ is a time-independent neutron flux function. Substituting from Equation 3.89 into the governing equation 3.88 yields the following relation for $\phi^*$:

$$\nabla^2\phi^* + \left\{\frac{(k_\infty - 1)\Sigma_a}{D} + \frac{\xi}{D\bar{u}}\right\}\phi^* = 0 \tag{3.90}$$

Note that, as in the steady state case, $\Sigma_a$ could be replaced using $\Sigma_a = D/L^2$ where $L$ is the neutron diffusion length (see the definition 3.19). In parallel with the steady state equation 3.24, Equation 3.90 can be written as the eigenequation

$$\nabla^2\phi^* + B_g^2\phi^* = 0 \tag{3.91}$$

where the geometric buckling, $B_g$, is the specific eigenvalue for the particular geometry of the reactor under consideration. Because Equation 3.91 is identical to that governing $\phi$ in the steady case, and because the boundary conditions are usually the same, the geometric buckling, $B_g$, will be the same as in the steady case. In addition, from Equations 3.90 and 3.91, it follows that

$$B_g^2 = \frac{(k_\infty - 1)}{L^2} + \frac{\xi}{D\bar{u}} \tag{3.92}$$

so that, using Equation 3.19,

$$\xi = D\bar{u}B_g^2 + \bar{u}\Sigma_a - k_\infty\bar{u}\Sigma_a \tag{3.93}$$

in which the left-hand side consists of contributions to $\xi$ from the neutron leakage, absorption, and production, respectively. Alternatively, the quantity $t^*$ can be defined by

$$\xi = \frac{(k - 1)}{t^*} \tag{3.94}$$

where $k$ is the multiplication factor and $t^*$ is the mean lifetime of a neutron in the reactor, where, using Equations 3.29 and 3.19,

$$k = \frac{k_\infty}{1 + L^2 B_g^2} \qquad t^* = \frac{1}{\bar{u}\Sigma_a(1 + L^2 B_g^2)} \tag{3.95}$$

Note that $(\bar{u}\Sigma_a)^{-1}$ is the typical time before absorption and $(\bar{u}\Sigma_a L^2 B_g^2)^{-1}$ is the typical time before escape; combining these, it follows that $t^*$ is the typical neutron lifetime in the reactor.

Hence the solution to the characteristic unsteady problem may be written as

$$\phi = C \exp\left(-\frac{(k-1)}{t^*}t\right)\phi^* \tag{3.96}$$

where $\phi^*$ is the solution to the steady diffusion problem with the same geometry and boundary conditions. The characteristic response time of the reactor, $t_R$, is known as the *reactor period*. In the absence of other factors, this analysis and Equation 3.96 suggest that $t_R$ might be given by

$$t_R = \frac{t^*}{(k-1)} \tag{3.97}$$

Because the typical lifetime of a neutron, $t^*$, in a LWR is of the order of $10^{-4}$ s, Equation 3.97 suggests that a very small perturbation in the multiplication factor $k$ of 0.1 percent to 1.001 might result in a reactor period, $t_R$, of 0.1 s and therefore more than a $2 \times 10^4$-fold increase in the neutron population in 1 s. This would make any reactor essentially impossible to control. Fortunately, as described in Section 2.3.4, delayed neutron emission causes a more than 100-fold increase in the mean neutron lifetime in an LWR and a corresponding increase in the reactor period, making reactor control much more manageable (see Section 4.3.6).

### 3.9.2 Point Kinetics Model

To incorporate the delayed neutrons in the analytical model and therefore allow modeling of the transients associated with practical reactor control, as described in Section 2.3.4, it is necessary to expand the preceding model to allow for the neutron flux associated with the delayed neutrons. To do this, the source term, $S$, in Equation 3.87 must be subdivided into contributions from the prompt neutrons and the delayed neutrons. If the fraction of delayed neutrons is denoted by $\beta$, then the contribution to $S$ from the prompt neutrons will be $(1-\beta)k_\infty\Sigma_a\phi$. The contribution from the delayed neutrons is normally modeled as the sum of contributions from each of the chosen precursor types (usually 6 in number), each with its own concentration $C_i$, $i = 1$ to 6, and decay constant, $\lambda_i$. Consequently, the diffusion equation 3.87 becomes

$$\frac{1}{\bar{u}}\frac{\partial\phi}{\partial t} - D\,\nabla^2\,\phi = \left\{(1-\beta)k_\infty\Sigma_a - 1\right\}\phi + \sum_{i=1}^{6}\lambda_i C_i \tag{3.98}$$

This modified diffusion equation along with a population equation for the concentration (see, e.g., Knief 1992) of each of the six precursor types constitute what is known as a *point kinetics model* for the dynamics of reactors. This type of model is essential for the realistic modeling of reactor dynamics.

### 3.10 More Advanced Neutronic Theory

As described in the earlier sections of this chapter, many approximations were made during the development of the diffusion theory described in the preceding. Consequently, though the one-speed diffusion theory results presented have qualitative instructional value, they are not adequate for practical reactor design. For this, much more detailed and complex analyses have been developed but are beyond the scope of this text. For further information on these more advanced calculational methods, the reader is referred to texts like Glasstone and Sesonske (1981) or Duderstadt and Hamilton (1976).

## 3.11 Monte Carlo Calculations

Before leaving the problem of determining the neutron flux in a reactor (and by extension the generation of heat within the core), it is appropriate to describe briefly an entirely different calculational procedure that is used increasingly and that, at least superficially, appears to bypass all the integro-differential equations of the preceding sections. These are known as Monte Carlo methods, and the simplest description of these is that one chooses to follow an individual neutron as it proceeds through a whole sequence of interactions within the core. Each interaction is governed by a selected probability distribution, and the outcome of the interaction is determined by a known or estimated probability distribution combined with a random number generator. Fission interactions determine the next generation of neutrons, their number as well as their speed, direction, and origin. The calculation proceeds until a steady state is reached, one that is independent of the location, speed, and direction of the starting neutron or neutrons. Throughout the calculation, the average neutron transport properties are assessed at every location within the reactor, and when these properties asymptote to a constant value (in a calculation that seeks a steady state), then, provided the result is independent of the initial neutron distribution and properties selected, a potential solution to the neutron flux distribution has been found.

As with many other multiphase flow calculations, it may still not be possible to simulate the kind of neutron or particle populations that are present in real reactors; in such cases, methods known as *multiscale* or *reduced-order* models have been developed in which a much smaller population is used to simulate a much greater population. The key with these reduced-order models is the determination of the interaction coefficients in the reduced-order model that are appropriate to modeling the interactions in the full-scale application.

Of course, many details within the numerical method require much more attention, but the rapid expansion in computing power and the ability to adapt these relatively simple calculations to parallel computers have meant that these methods have become increasingly used and useful. For further detail, the reader is referred to texts such as Carter and Cashwell (1975).

### REFERENCES

Bell, G. I., and Glasstone, S. (1970). *Nuclear reactor engineering.* Van Nostrand Reinhold.
Carter, L. L., and Cashwell, E. D. (1975). Particle transport simulation with the Monte Carlo method. *U.S. Energy Research and Development Administration* Rep. TID-26607.

Duderstadt, J. J., and Hamilton, L. J. (1976). *Nuclear reactor analysis.* John Wiley.

Glasstone, S. (1955). *Principles of nuclear reactor engineering.* Van Nostrand.

Glasstone, S., and Sesonske, A. (1981). *Nuclear reactor engineering.* Van Nostrand Reinhold. [See also Glasstone (1955) and Bell and Glasstone (1970)]

Knief, R. A. (1992). *Nuclear engineering: Theory and practice of commercial nuclear power.* Hemisphere.

# 4

# Some Reactor Designs

## 4.1 Introduction

Discussions of current and future nuclear reactor designs utilize a convenient international notation based on the decades in which the designs originated. Thus Generation I reactors refer to early prototype reactors designed and built in the 1950s and early 1960s. Generation II reactors are those designed and built in the 1970s and 1980s and therefore include most of the commercial power generating reactors in operation today. Generation III reactors from the late 1990s, 2000s, and 2010s are few in number and can be characterized as advanced LWRs, evolutionary designs offering improved economics. Generation IV reactors refer to those that might be possible, given the focus of current research and development exploration. This chapter devoted to reactor designs focuses primarily on Generations II and IV.

As illustrated in Figure 4.1, a nuclear power plant is similar to any other coal, gas, or oil-fired plant, except that the source of the heat creating the steam that drives the steam turbines and therefore the electrical generators is the nuclear reactor core rather than the fossil fuel furnace. The focus in this text is on that core, known as the *nuclear steam supply system* or *NSSS*. It is assumed that the reader is familiar with the rest of the equipment (known as the *balance of plant*).

Before proceeding with further analysis, it is useful to provide some engineering context by briefly describing the design and components of current Generation II reactors. Consequently, the focus of the first part of this chapter is on Generation II reactors and those designs used in commercial power generating reactors in operation today. The last sections briefly describe some of the ideas being explored as Generation IV reactors.

## 4.2 Current Nuclear Reactors

Typical data on some of the principal types of Generation II reactors are listed in Table 4.1. These differ primarily in terms of the nuclear fuel being utilized and therefore the nuclear fuel cycle involved (see Section 2.2), and this distinguishes the LWRs from the HWRs and the FBRs. The two LWR types are then distinguished by the strategy used to handle the possibility of the cooling water boiling and therefore

Table 4.1. *Some typical nuclear power plant data for Generation II reactors (Figure 4.2)*

| Type of reactor | PWR (LWR) | BWR (LWR) | CANDU (HWR) | LMFBR (FNR) |
|---|---|---|---|---|
| Electrical output (MW) | 1150–1300 | 1200 | 500 | 1000 |
| Efficiency (%) | 33–34 | 33 | 31 | 39 |
| Fuel | $U_2O$ | $U_2O$ | $U_2O$ | $U_2O$, $PuO_2$ |
| Primary coolant | $H_2O$ | $H_2O$ | $D_2O$ | Na |
| Moderator | $H_2O$ | $H_2O$ | $D_2O$ | None |
| Coolant pressure (atm) | 155 | 72 | 89 | 1.4 |
| Coolant inlet (°C) | 296–300 | 269 | 249 | 380 |
| Coolant outlet (°C) | 328–333 | 286 | 293 | 552 |
| Flow rate ($10^6$ kg/hr) | 65 | 47 | 24 | 50 |
| Max. fuel temp. (°C) | 1788–2021 | 1829 | 1500 | 2000 |

*Source:* Data are extracted from Duderstadt and Hamilton (1976) and Todres and Kazimi (1990).

by the pressure of the primary cooling water system and the corresponding safety systems. Each of these features is a focus in the sections that follow.

## 4.3  Light Water Reactors (LWRs)

### 4.3.1  Types of LWRs

By far the greatest fraction of nuclear reactors used to produce power around the world belong to the class known as light water reactors (LWRs), in other words,

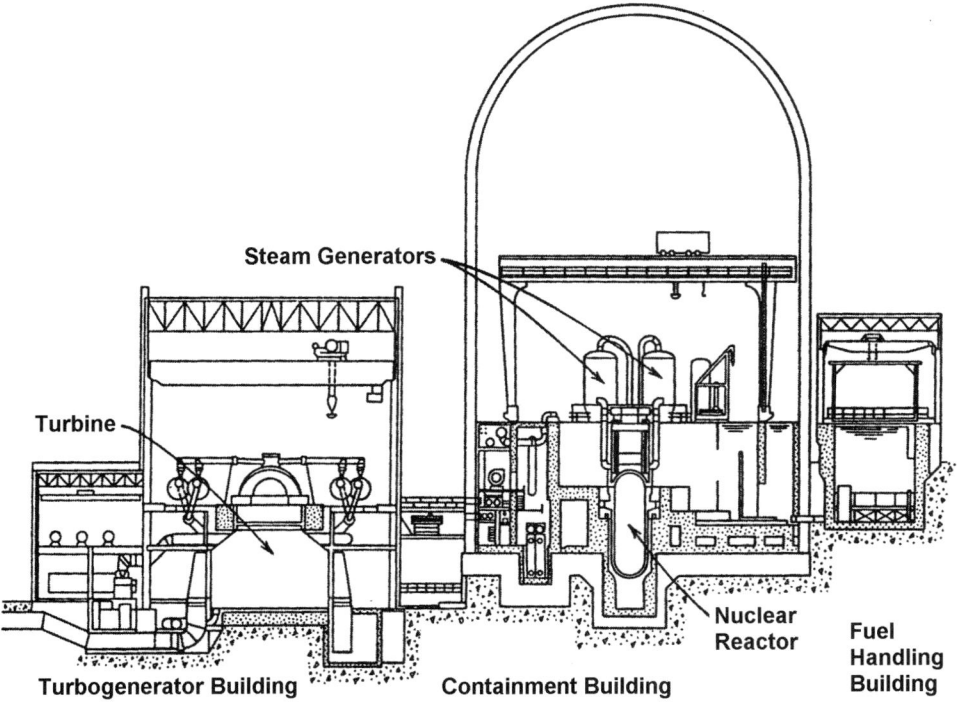

Figure 4.1. Schematic of a nuclear power plant. From Duderstadt and Hamilton (1976).

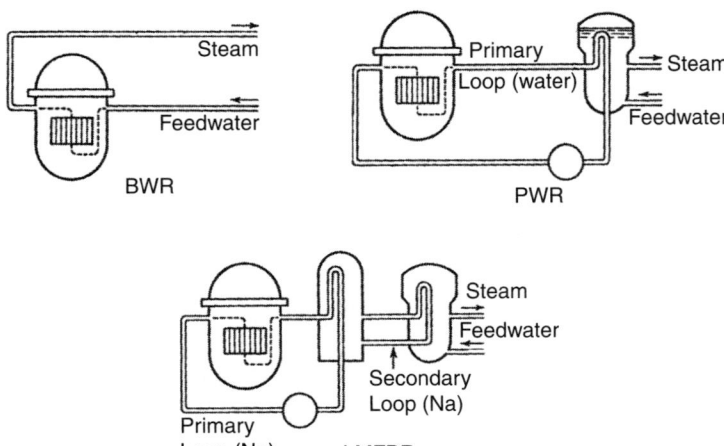

Figure 4.2. Schematics of a boiling water reactor, BWR, a pressurized water reactor, PWR, and a liquid metal fast breeder reactor, LMFBR. From Duderstadt and Hamilton (1976).

reactors that utilize *light* water (as opposed to *heavy* water) as the moderator and primary coolant. To be self-sustaining neutronically, a LWR with natural uranium fuel must use heavy water as the moderator to maintain the neutron flux. The Canadian-designed CANDU heavy water reactor operates on this basis and is described in more detail in Section 4.8. LWRs, conversely, require enriched uranium fuel to be self-sustaining. However, because light water absorbs neutrons as well as slowing them down, it is less efficient as a moderator than heavy water or graphite.

Besides serving as both moderator and primary coolant, water has many advantages in this context. It is inexpensive, and the technology of water cooling is very well known and tested; it also has a high heat capacity and a low viscosity so that the heat can be removed with relatively low flow rates and pressure drops. Burnable poisons that absorb neutrons are often added to the primary coolant water to provide some additional control over the reactivity and to even it out over time. Most importantly, in most (though not all) designs of LWRs, boiling of the water within the core leads to a decrease of reactivity and serves as an automatic reactor shutdown mechanism (see Section 7.4).

Various types of light water reactors have been developed in the past decades. These can be subdivided into two principal types, namely, pressurized water reactors (PWRs) and boiling water reactors (BWRs), which are described in Sections 4.3.2 and 4.3.3.

### 4.3.2 Pressurized Water Reactors (PWRs)

The majority of light water reactors (LWRs) in operation in the world are known as *pressurized water reactors* (PWRs) because water is used to remove the heat from the core and because the primary coolant loop is pressurized to suppress boiling. In 2013, approximately 270 of these were in commercial operation worldwide. An acceptably large thermodynamic efficiency is only achieved by having a primary cooling system

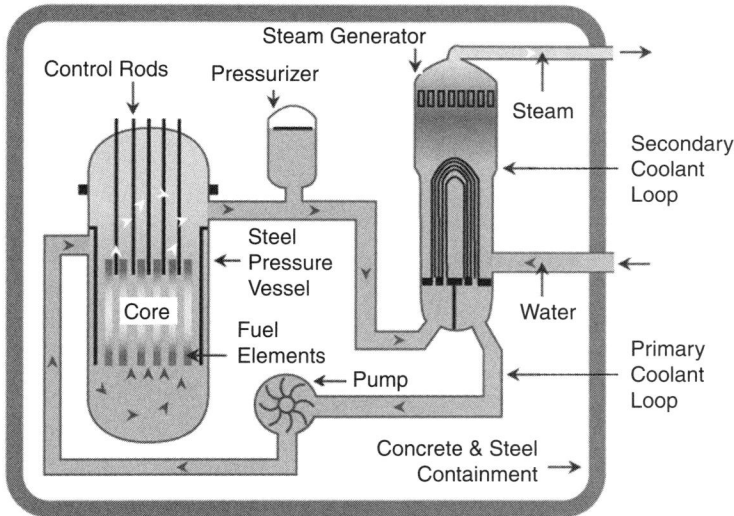

Figure 4.3. Schematic of a typical PWR. Adapted from WNA (2015b).

that operates at a high maximum temperature, and these high temperatures would result in boiling unless that primary coolant loop were pressurized. The alternative would be to allow boiling and to remove most of the heat from the core in the form of latent heat; that alternative strategy is followed in the other major design, namely, the boiling water reactors (or BWRs) that are covered in the section that follows.

A schematic of the typical PWR is illustrated in Figure 4.3 and includes a reactor vessel such as that cross sectioned in Figure 4.4 and equipped with a primary coolant system like that of Figure 4.5. All of this and more is contained in a containment building such as that shown in Figure 7.3 and described later in Section 7.5.1 (see also USNRC 1975). The primary coolant inlet and outlet temperatures (from the reactor vessel) are approximately 300°C and 330°C, respectively, but with the high specific heat of water, this modest temperature difference is adequate to transport the heat at reasonable water flow rates of the order of $65 \times 10^6$ kg/hr. However, to avoid boiling at these temperatures, the pressure in the primary coolant loop is 155 atm; this is maintained by pressurizers (see Figure 4.5) contained within the containment structure (Figure 7.3). The high pressure makes for a compact reactor with a high power density. However, the high pressure is also a liability in an accident scenario, and therefore this primary coolant loop is secured inside a heavy and strong containment building. A secondary coolant loop that operates at much lower pressure and is less susceptible to radioactive contamination communicates thermally with the primary loop in a heat exchanger and steam generator (Figures 4.5 and 7.3) within the containment building. The steam thus generated moves the heat outside of that building and is used to drive the steam turbines and electrical generators.

While this double coolant loop system involves some thermal inefficiency and some added equipment, it has the advantage of confining the high-pressure coolant water (and the radioactivity it contains) within the containment building. The building also houses extensive safety equipment that is described later in Section 7.4.

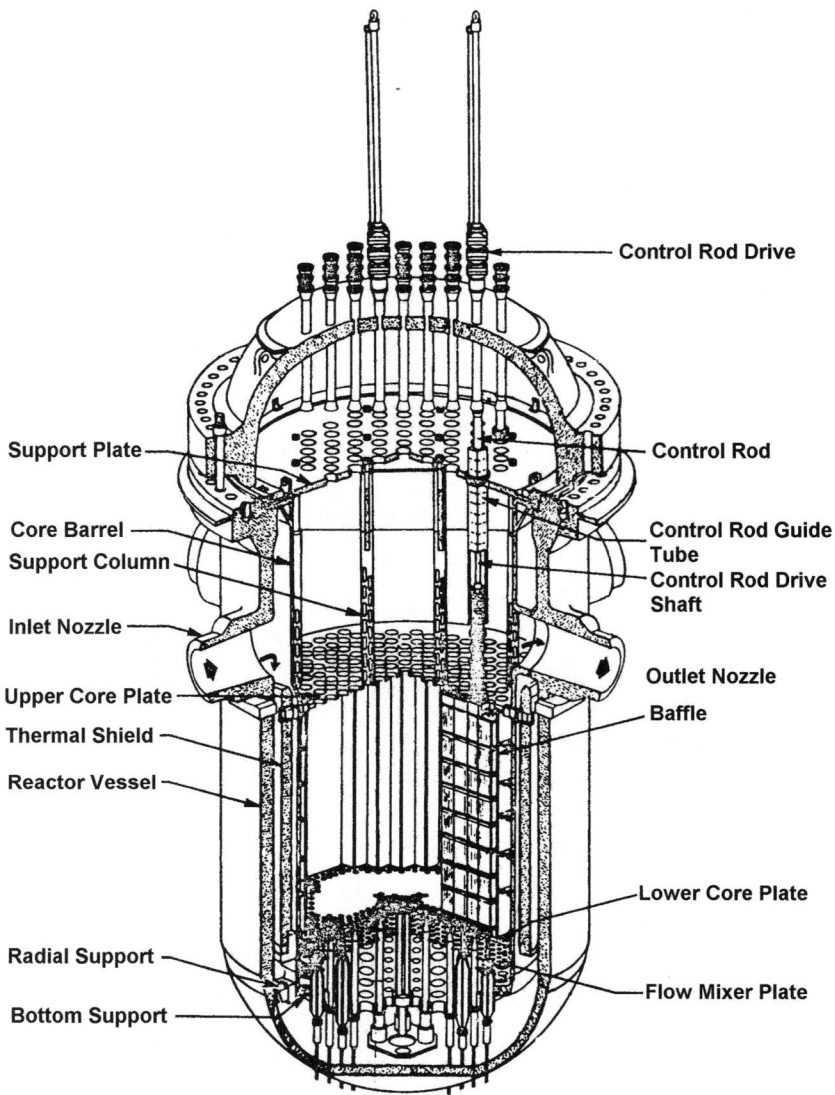

Figure 4.4. Internals of a typical PWR reactor vessel. Adapted from USAEC (1973).

### 4.3.3 Boiling Water Reactors (BWRs)

The concept behind the boiling water reactor (in 2013, approximately 84 of these were in commercial operation worldwide) is to avoid the high pressures of PWRs (and thus the associated dangers) by allowing the primary coolant water to boil as it progresses through the reactor core, as shown in the schematic of a typical BWR in Figure 4.6. As depicted in Figure 4.7, the steam thus generated is fed directly to the turbines, thus eliminating the secondary coolant loop. Details of the reactor core of a BWR are shown in Figure 4.8. By avoiding the high primary coolant loop pressures, this design reduces the need for the kind of large and costly containment structure deployed around a PWR (Figure 7.3) because a rupture in the primary coolant loop would not lead to such a high buildup of pressure inside that secondary containment.

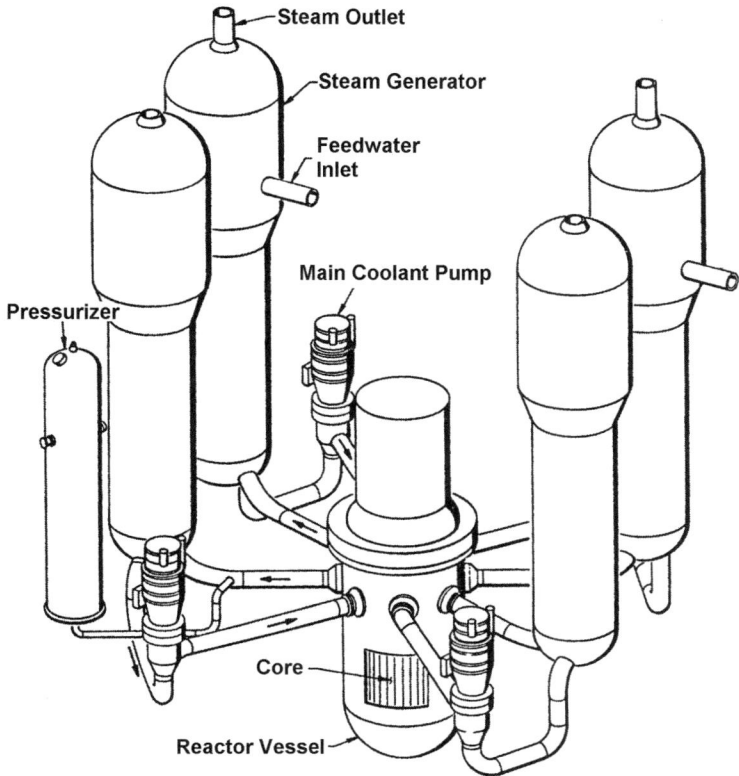

Figure 4.5. PWR coolant system. Adapted from USAEC (1973).

Instead, General Electric, which designed and built the BWRs, devised a secondary containment structure that, in the event of a primary coolant loop rupture, would direct the steam down through pipes into a large body of water (known as a suppression pool), where it would be condensed. This would minimize the buildup of steam

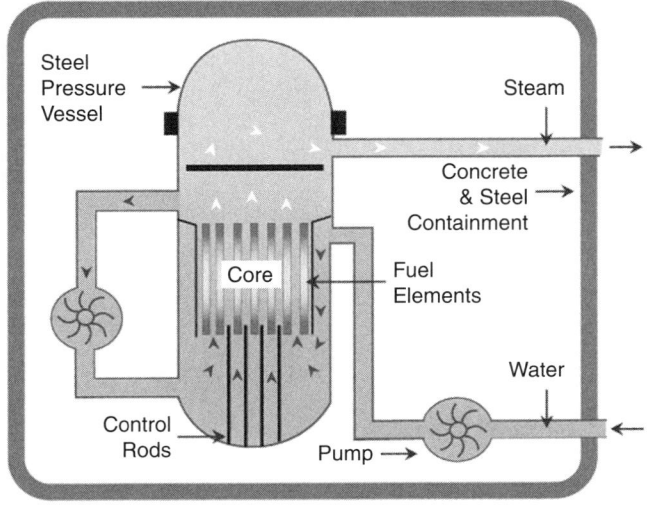

Figure 4.6. Schematic of a typical BWR. Adapted from WNA (2015b).

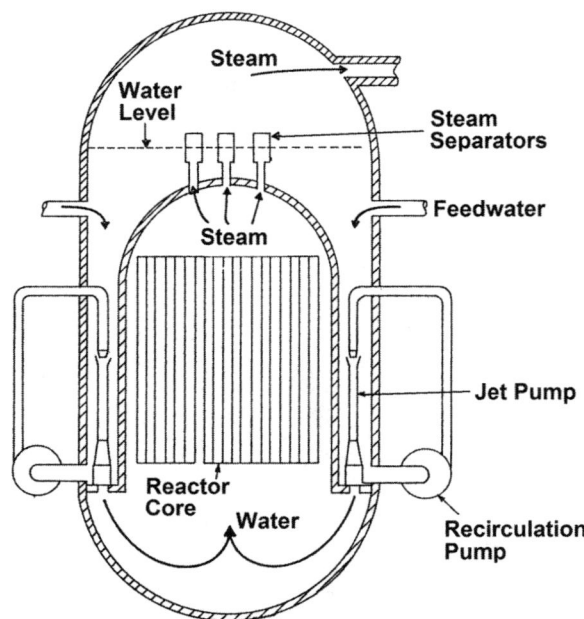

Figure 4.7. Schematic of the BWR coolant and steam supply systems. Adapted from USAEC (1973).

pressure within the secondary containment. The first (or Mark I) suppression pool was toroidal in shape, as shown later in Figure 7.5. Later, several other pressure suppression configurations were produced. Further comment on the issues associated with primary coolant loop rupture in a BWR are delayed until later (Section 7.4).

The elimination of the secondary or intermediate coolant loops is advantageous for the thermal efficiency of the unit, but it also means increased buildup of radioactivity in the turbines. Other features of the BWR include the effect of the steam–water mixture on the moderator role played by the coolant (see Section 7.4 on reactor control).

### 4.3.4 Fuel and Control Rods for LWRs

The uranium dioxide fuel in a PWR or BWR is formed into cylindrical pellets that are packed into zircaloy tubes about 3.5 m in length known as fuel rods (Figure 4.9). The walls of the fuel tubes are known as *cladding*. In a typical PWR, the pellets are 0.97 cm in diameter and the fuel rods have an outside diameter of 1.07 cm; in a typical BWR, the corresponding diameters are 1.24 cm and 1.43 cm, respectively. Typically the core contains 55,000 and 47,000 fuel rods, respectively, in a PWR and BWR. As seen in Section 5.3, these dimensions imply cylindrical reactor core dimensions (height and diameter) of about 3.6 m and 4.4 m, respectively.

The fuel rods are arranged in *fuel assemblies* or *fuel bundles*, as shown in Figure 4.9. In a PWR, the typical arrangement in a fuel assembly consists of a square cross-sectioned cell (Figure 4.10, left) containing approximately 200 equally spaced fuel

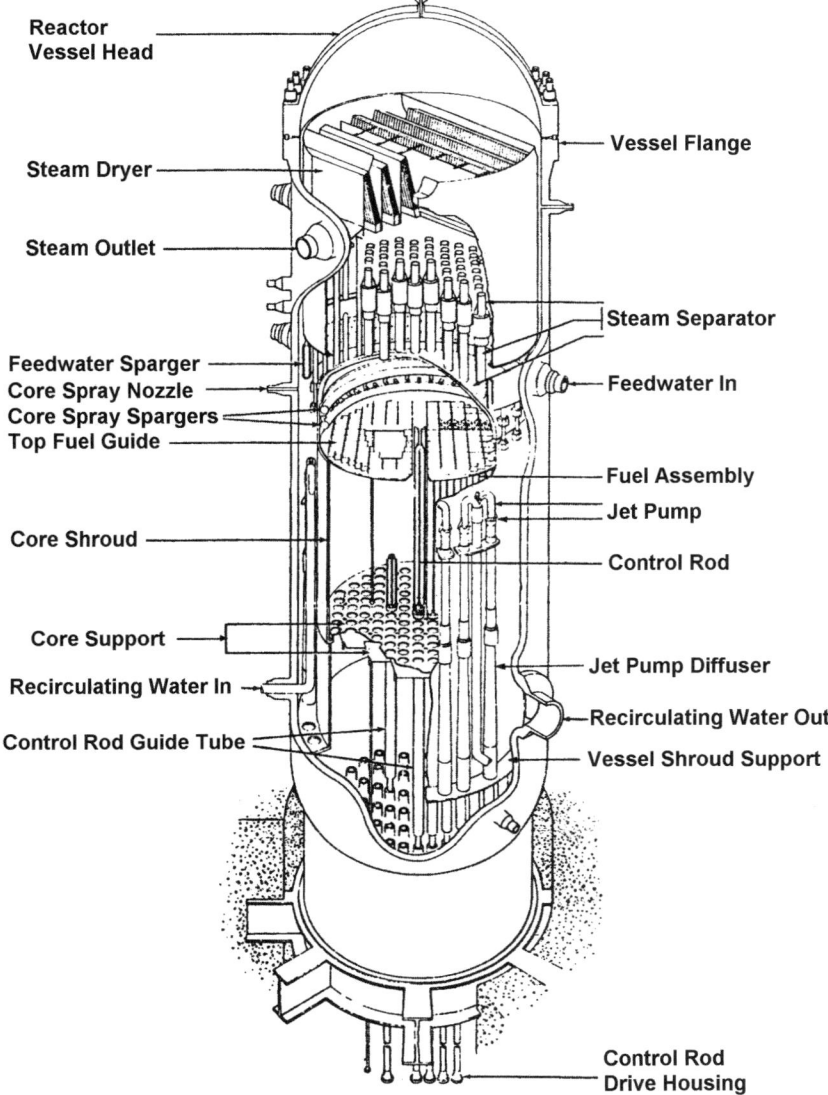

Figure 4.8. Typical BWR reactor vessel. Adapted from USAEC (1973).

rods interspersed with approximately 20 circular control rod channels; the coolant in the cell flows in the spaces between these elements. There are approximately 200 of these assemblies arranged, lattice-like, in a PWR core. A BWR core also consists of cells (Figure 4.10, right) each containing about 64 fuel rods arranged in a square channel through which the coolant flows. Four of these cells are grouped together with the rectilinear space between them containing the cruciform-shaped control blade. There are approximately 180 such groups of four assemblies in a BWR core.

Thus the fuel rods, control rods, moderator, coolant channels, and so on, in a reactor are usually arranged in *lattice cells* that are repeated across the cross section

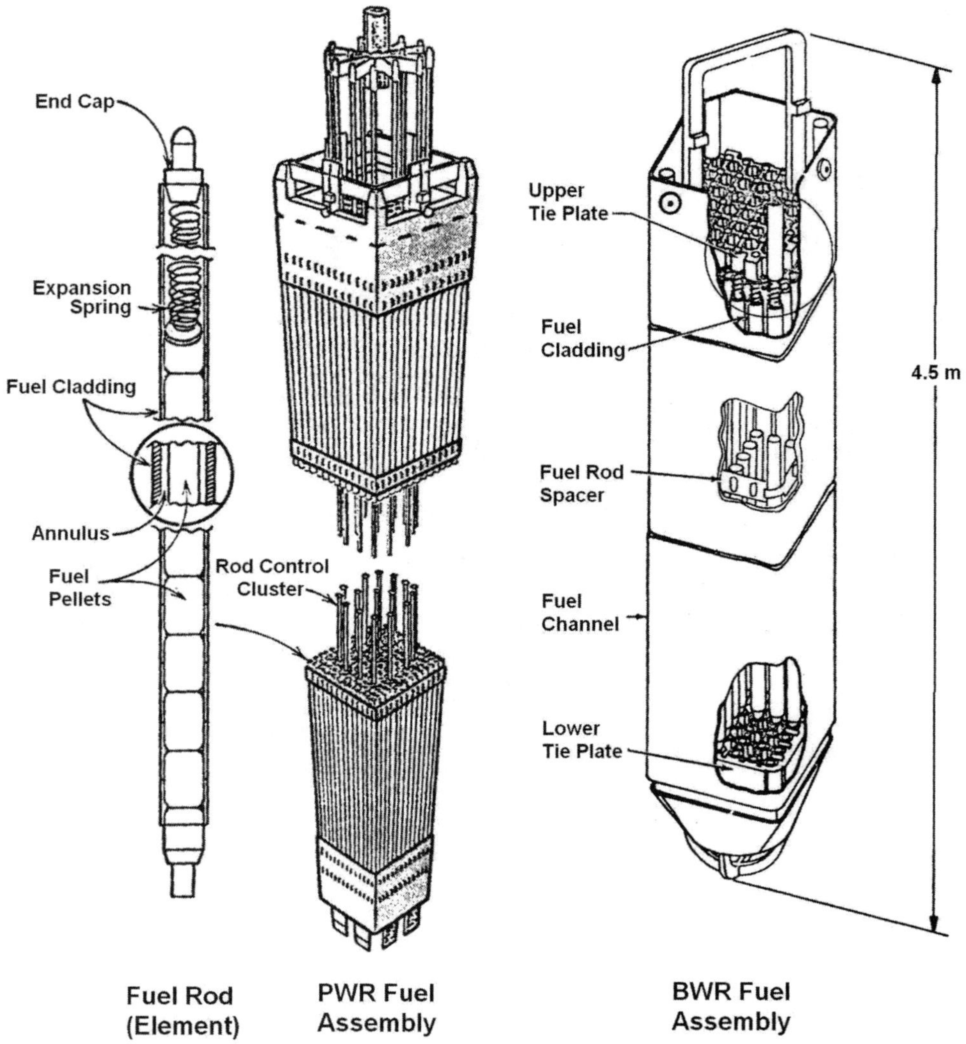

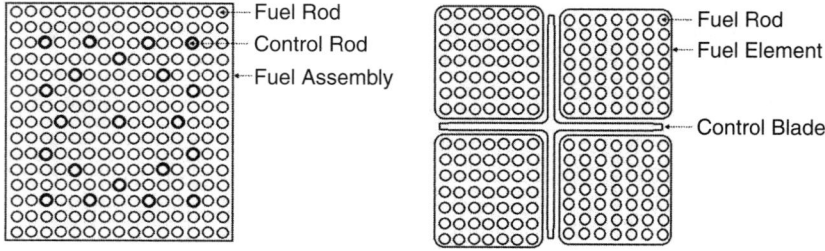

Figure 4.9. Fuel element and PWR fuel assembly (from Duderstadt and Hamilton 1976) and BWR fuel assembly (from USAEC 1973).

of the core. Consequently, there are several structural or material scales within the core, and these various scales of inhomogeneity become important in some of the more detailed calculations of the neutron flux within the reactor (see Section 3.6.7).

Figure 4.10. Cross sections of PWR (left) and BWR (right) fuel assemblies.

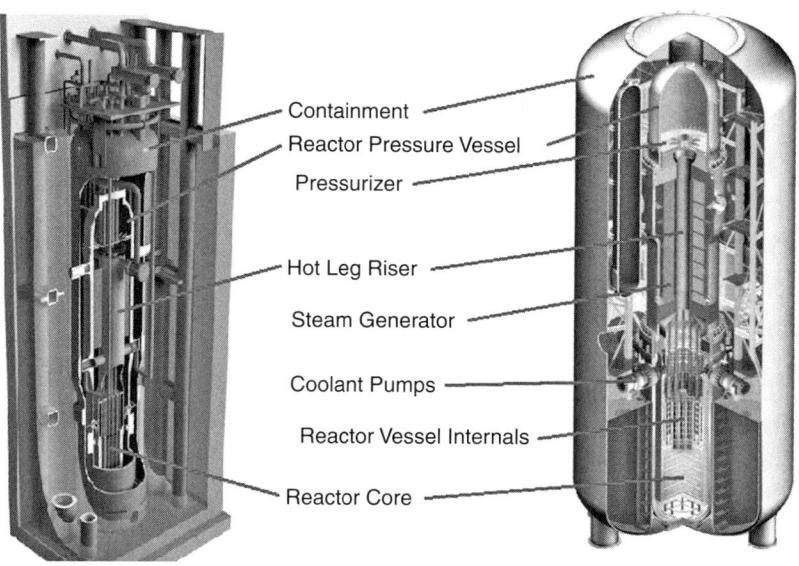

Containment
Reactor Pressure Vessel
Pressurizer

Hot Leg Riser
Steam Generator
Coolant Pumps
Reactor Vessel Internals
Reactor Core

Figure 4.11. Sketches of the NuScale (left) and Westinghouse (right) SMRs (not to scale).

### 4.3.5 Small Modular Reactors

In recent years, there has been an increase of interest in the development of small modular reactors (SMRs) about a quarter of the size of conventional large-scale commercial reactors. Most (though not all) of the SMR designs are traditional pressurized water reactors designed to be manufactured in a factory and transported as a whole to their operating location. This would allow faster construction and more flexibility in deploying and ultimately disposing of the reactor. Examples shown in Figure 4.11 include the proposed Westinghouse SMR that generates 225 MW of electricity; conceptually 25 of these reactor containment vessels units would fit within the containment of a full-scale Westinghouse AP1000 reactor plant. The 27-m-tall containment vessel encloses a 24.7-m-tall reactor vessel housing a core with 89 assemblies loaded with < 5 percent enriched $U^{235}$. The passive safety systems require no operator intervention for 7 days. Also shown is the smaller NuScale SMR designed to generate 45 MW of electricity; the 24.3-m-tall containment enclosing a 19.8-m-tall reactor vessel operates in a below-grade, water-filled pool of water. It is designed to safely shut down and self-cool indefinitely with no operator action, no electric power, and no additional water. These SMRs are also designed to be combined in a multiple reactor array with a unified control and safety system. In a general sense, they reflect an operational strategy similar to that of the small gas turbine generating units, designed for flexibility in deployment and usage as well as speed of construction. Thus some of the decrease in efficiency is offset by the reduced start-up and shutdown durations.

It is interesting to reflect that, in the early days of nuclear reactor development, Aerojet-General advertised a personal nuclear reactor, the AGN201, that is depicted in the advertising postcard shown in Figure 4.12. The blurb on the back of the postcard is particularly disingenuous and illustrates how much public perceptions of reactor safety have changed.

Figure 4.12. Postcard advertising a personal reactor, circa 1950: front (top); obverse (bottom).

### 4.3.6 LWR Control

The need to maintain tight control on the operation of a nuclear reactor is self-evident, and this control is maintained using a variety of tools, managerial, mechanical, and chemical. In Section 3.9, it was observed that control was made much easier, indeed, one might say made practical, by the delayed neutrons that extend the neutronic response time of the reactor core by several orders of magnitude. Indeed, if the neutron population consisted only of prompt neutrons, the calculations of Section 3.9 demonstrate that the reactor control system would have to respond in fractions of a second to maintain control. The presence of delayed neutrons allows response times of the order of tens or hundreds of seconds to maintain control. The corollary is that the prompt neutron population of a reactor must always be maintained well below the critical level in all sections of the reactor core and throughout the history of the fuel load. It is the delayed neutrons that are used to reach criticality and are manipulated to increase or decrease the power level.

The primary mechanical devices that are used to effect control are the control rods (or structures) that are inserted into channels in the core, as described in the preceding section. These are fabricated from material that absorbs neutrons and, when inserted, decreases the reactivity of the core. The materials used include boron, cadmium, and gadolinium. As indicated in Figure 4.9, the control rods are usually motor driven from above and sometimes set to drop into the core without power in emergency situations. A full control rod insertion under emergency conditions is referred to as a *scram* and the process as *scram control*. The control rods are also used to adjust the power output from the reactor and to compensate for the aging of the fuel over longer periods of time (known as *shim control*). Typically, a LWR is initially loaded with enough fuel to achieve a multiplication factor, $k$ (see Section 2.3.1), of as much as 1.25, and therefore sufficient control rod insertion is needed to balance the reactor. As fuel life is expended, the insertion is correspondingly decreased.

In addition to the control rods, several other methods are used to adjust the power level of the reactor, to compensate for the aging of the fuel and to balance the power produced in different regions of the core. Absorbing materials are sometimes fixed in the core to age with the fuel and even out the long-term power production. Another strategy is to dissolve absorbing or *burnable* poison, such as boric acid, in the coolant.

## 4.4  Heavy Water Reactors (HWRs)

An alternative thermal reactor design that uses natural rather than enriched uranium is the heavy water reactor (HWR). The principal representative of this class of reactors is the Canadian-built CANDU reactor (see, e.g., Cameron 1982; Collier and Hewitt 1987), of which there were about 48 in commercial operation worldwide as of 2013. A schematic of the CANDU reactor is included in Figure 4.13. The use of natural uranium fuel avoids the expense of the enrichment process. In an HWR, the reactivity is maintained by using heavy water ($D_2O$) rather than light water as the moderator.

One of the unique features of the CANDU reactor is the refueling technique employed that is made possible by the natural uranium fuel. As depicted in Figure 4.13, the fuel is contained in horizontal tubes and refueling is done continuously rather than in the batch process used in LWRs. Fueling machines inside the secondary containment push the natural uranium *fuel bundles* into the core and remove the spent fuel bundles at the other side of the reactor. The coolant, instead of being contained in a primary pressure vessel, as in the LWRs, flows through the core in horizontal pressure tubes surrounding the fuel channels, of which there are typically 380–480 in a CANDU reactor.

The cylindrical fuel bundles that are pushed through the core in the fuel channels are about 10 cm in diameter and 50 cm long. They consist of a zircaloy package of about 30–40 zircaloy fuel tubes that contain the fuel in pellet form. In an older

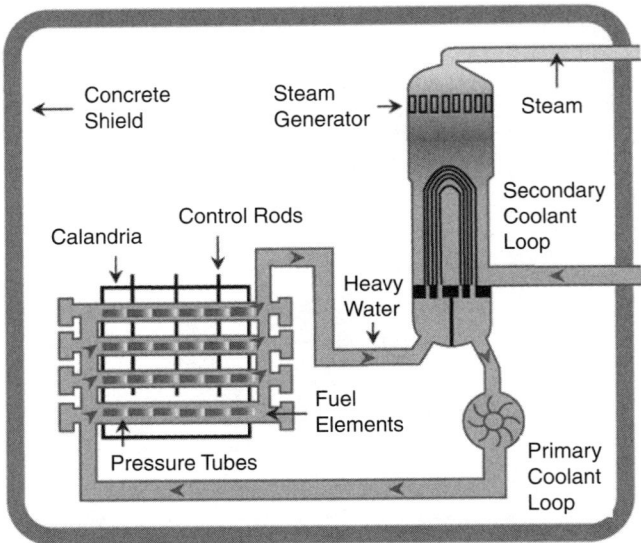

Figure 4.13. Schematic of the CANDU heavy water reactor. Adapted from WNA (2015b).

model, 12 of these fuel bundles lay end-to-end within each fuel channel. Light water coolant flows through high-pressure tubes surrounding the fuel channels, and these high-pressure coolant tubes are in turn surrounded by a *calandria tube* containing a thermally insulating flow of carbon dioxide gas. All of this tube assembly is contained in a much larger, low-pressure tank known as the *calandria* that contains most of the heavy water moderator. The carbon dioxide flow placed between the light water coolant and the heavy water moderator is needed to prevent the hot coolant from boiling the moderator. Note that a cooling system is also needed for the heavy water moderator; this moderator mass represents a heat sink that provides an additional safety feature.

As described in Section 2.8.1, the heavy water moderator is needed with natural uranium fuel because the heavy water absorbs a lesser fraction of the neutrons and thus allows a sustainable chain reaction. However, a larger presence of heavy water moderator is needed to slow the neutrons down to thermal energies (because the heavier deuterium molecule needs more collisions to slow down the neutrons), and therefore the CANDU reactor requires a larger thickness of moderator between the fuel bundles. This means a proportionately larger reactor core.

One of the disadvantages of the CANDU reactor is that it has a positive void coefficient (see Section 7.1.2). In other words, steam formed by coolant boiling would cause an increase in the reactivity that, in turn, would generate more steam. However, the much larger and much cooler mass of moderator in the calandria would mitigate any potential disassembly. Other features of the design that improve the margin of safety include the basic fact that natural uranium fuel is not critical in the light water coolant and the fact that any distortion of the fuel bundles tends to reduce the reactivity. The CANDU reactor also contains a number of active and passive safety features. As well as the normal control rods, shut-off emergency control

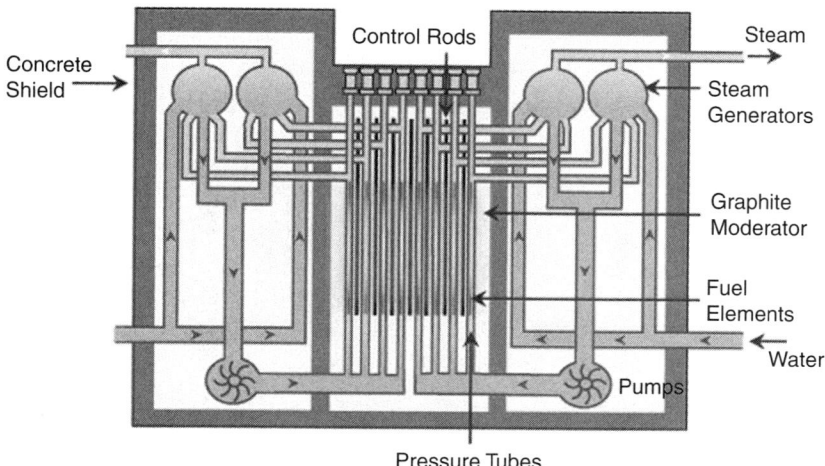

Figure 4.14. Schematic of the Chernobyl RBMK boiling water reactor. Adapted from WNA (2015a).

rods are held above the core by electromagnets and drop into the core, if needed. Another high-pressure safety system injects a neutron absorber into the calandria in the event of an emergency.

## 4.5 Graphite-Moderated Reactors

One of the older Russian designs, notorious because of the Chernobyl disaster, is the enriched uranium, water-cooled BWR known by the initials RBMK. This is shown schematically in Figure 4.14. More than 10 of these are still in commercial operation worldwide, though substantial modifications have been made since the disaster. For moderator, these reactors utilize graphite as well as the coolant water and have the severe disadvantage that additional boiling within the core does not necessarily lead to a decrease in reactivity. Rather, the reactivity can increase as a result of a loss of coolant, and this may have been a factor in the Chernobyl accident (see Section 7.5.2).

## 4.6 Gas-Cooled Reactors

Yet another alternative is the gas-cooled reactor design (see, e.g., Gregg King 1964). Some 17 of these were, as of 2013, in commercial operation (mostly in the United Kingdom), cooled by $CO_2$ and moderated by graphite. Early versions (now superseded) utilized natural uranium, though this required large cores. The more recent, advanced gas reactors (AGR) use enriched uranium as fuel. Their design is shown conceptually in Figure 4.15 (WNA 2015b; Winterton 1981). The $CO_2$ flows up through channels in the bricks of the graphite moderator. These channels are interspersed with control rod channels. The entire core is surrounded by a thermal shield, and the $CO_2$ flow loop passes up the outside of the shield and down its inside before entering the bottom of the core. Heat exchanger-steam generator tubes to transfer

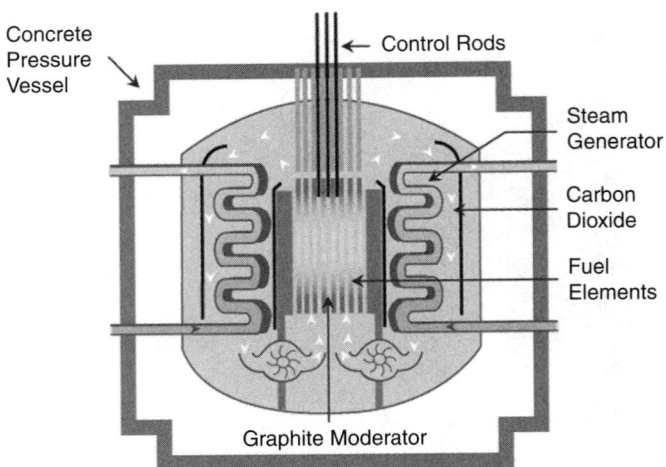

Figure 4.15. Schematic of the typical advanced gas reactor. Adapted from WNA (2015b).

the heat to the secondary water coolant circuit are enclosed with the core in the primary containment structure, a pre-stressed concrete vessel.

Note should also be made of the more recently proposed design in the United States, the high-temperature gas-cooled reactor (HTGR) that utilizes high-pressure helium as the coolant (Duderstadt and Hamilton 1976). This design has a quite different fuel cycle with an initial reactor core loading of highly enriched uranium carbide along with thorium oxide or carbide and graphite moderator. The design has the advantage of more efficient use of the uranium, though whether it will be used for power generation remains to be seen.

## 4.7 Fast Neutron Reactors (FNRs)

As described in Section 2.9, the label *fast neutron reactor* (FNR) refers to a broad class of reactors that rely on fast neutrons alone to sustain the chain reaction. Consequently, there is no moderator. Various fuels and combinations of fuels can provide the required self-sustaining reaction. However, they are most often fueled with plutonium or a mixture of uranium and plutonium. Because there is a large store of highly enriched uranium that has been produced for military purposes, this is sometimes added to the fuel of fast reactors.

Often the core of a fast reactor is surrounded by a *blanket* of fertile $^{238}$U, in which the neutron flux from the central core produces or breeds additional plutonium; indeed, in the presently constructed fast breeder reactors (FBRs), most of the $^{239}$Pu is produced in this blanket.

## 4.8 Liquid Metal Fast Breeder Reactors

Because their power density is significantly higher than LWRs, the FBRs that have been constructed have been cooled by liquid metal, as the moderator effect of water is unwanted and liquid metals have a low moderating effect. Moreover, liquid metals have the advantage that they have a high thermal conductivity and can be operated at low pressures. This avoids the dangers that are associated with the high pressures in water-cooled reactors. Despite this, substantial safety issues are associated with FBRs that are addressed in Section 7.7 and that have limited their deployment to date. Nevertheless, some 20 LMFBRs in the world are currently producing electricity, and many more proposals have been put forward (see Section 4.9.2).

Sodium has been the universal choice for the primary coolant in LMFBRs for several reasons (lithium is another possibility though it has been, as yet, unused). First, sodium has high thermal conductivity, making it a good coolant, even though its heat capacity is about one-third that of water. Typically the primary coolant loop or pool functions at elevated temperatures of 395°C–545°C to achieve high thermal efficiency, but the pressure this requires is low (order of 0.1 MPa) because these temperatures are well below the boiling point of sodium at normal pressures (883°C). Thus most LMFBRs have a primary coolant loop or pool pressure just slightly above atmospheric, and this feature has significant safety advantages. Of course, the violent reactions of sodium with air and water require a very tight coolant loop system and some well-designed safety systems. Also, with a low atomic weight of 23, the scattering cross section for sodium is small, and therefore the neutron loss due to slowing is limited. Sodium also becomes radioactive when bombarded with neutrons, and so the primary coolant loop must be confined within a containment system and the heat removed by means of a heat exchanger and a secondary coolant loop. This secondary loop also uses liquid sodium but does not have the radioactivity of the primary coolant.

Two types of LMFBRs have been designed and constructed, the distinction being the configuration of the primary coolant loop. The so-called loop-type and pool-type LMFBRs are sketched diagrammatically in Figure 4.16. In the loop-type LMFBR, the primary coolant is circulated through the core by a primary coolant pump in the conventional way. Because of the high radioactivity, all these components require substantial shielding. These shielding requirements are significantly simplified in the other pool-type reactor, in which the core is submerged in a pool of sodium that is part of the primary coolant loop, and this pool as well as the heat exchanger to the secondary coolant loop are all enclosed in a large containment vessel. The Russian BN-600 reactor (Figure 4.17) and the French Phenix reactors (Figure 4.18) are both examples of pool-type LMFBRs.

In most LMFBRs, the fuel rods consist of stainless steel tubes about 0.6 cm in diameter containing the fuel pellets of oxides of uranium and plutonium. The rods are held apart by spacers and packed in fuel assemblies contained in stainless steel cans about 7.6 cm across and 4.3 m long. There are typically 217 fuel rods in each assembly and 394 assemblies in a reactor core. To achieve higher packing densities for the fuel rods, fast reactor fuel assemblies are always hexagonal, with the fuel rods in a triangular array, unlike the square arrangements in LWRs.

Arranged around the periphery of the core are the *blanket* fuel rods, which contain only uranium dioxide. Such a design creates a central *driver* section in the core

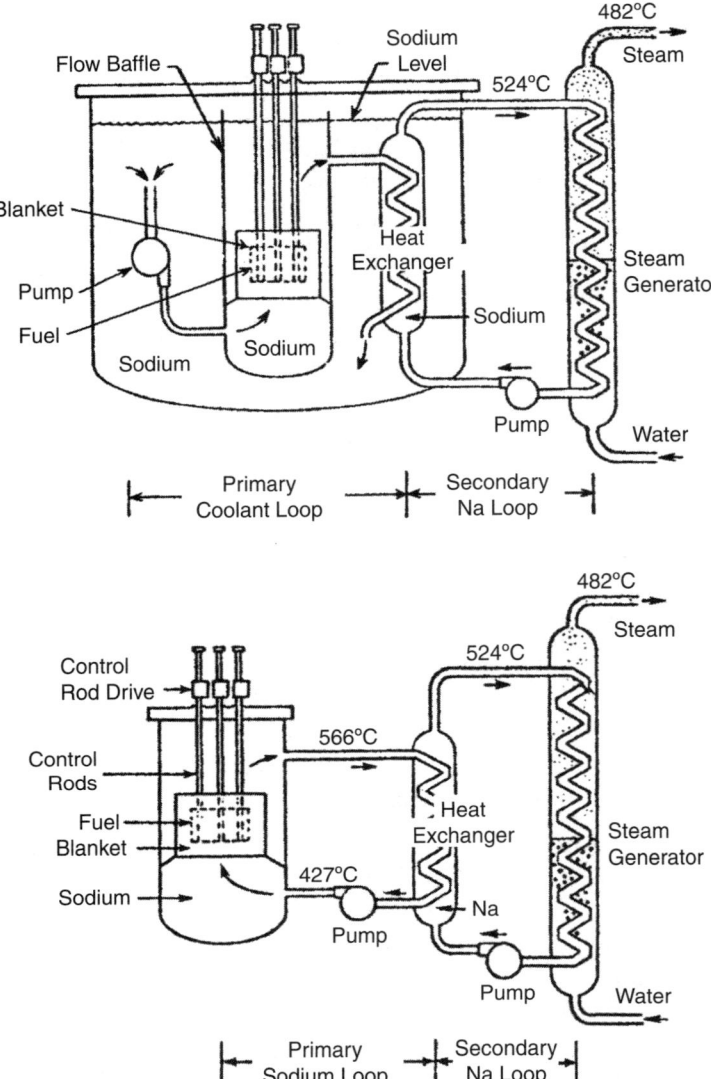

Figure 4.16. Schematics of a pool-type and a loop-type liquid metal fast breeder reactor. From Argonne National Laboratory, adapted from Wilson (1977).

surrounded on all sides by the blanket whose primary purpose is the breeding of new plutonium fuel (see Section 4.7). The core is quite small compared to a LWR core, measuring about 90 cm high and 220 cm in diameter for a core volume of 6.3 m³. It therefore has an *equivalent* cylindrical diameter and height of about 2.0 m (these reactor dimensions are commented on in Section 5.4). The flow pattern is similar to that of a PWR core in that the coolant flows upward though the core assembly and exits through the top of the core.

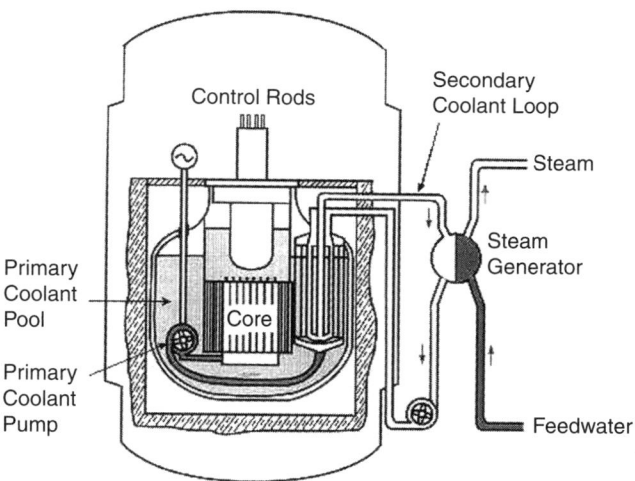

Figure 4.17. Schematic of Russian BN-600 pool-type LMFBR.

The BN-600 (Figure 4.17) is a Russian, pool-type, liquid sodium – cooled LMFBR that has been generating 600 MW of electricity since 1980 and was as of 2013 the largest operating fast breeder reactor in the world. The core (about 1 m tall with a diameter of about 2 m) has 369, vertically mounted fuel assemblies each containing 127 fuel rods with uranium enriched to 17–26 percent. The control and shutdown system utilizes a variety of control rods, and the entire primary coolant vessel with its emergency cooling system is contained in a heavily reinforced concrete containment building. The primary sodium cooling loop proceeds through a heat exchanger, transferring the heat to a secondary sodium loop that, in turn,

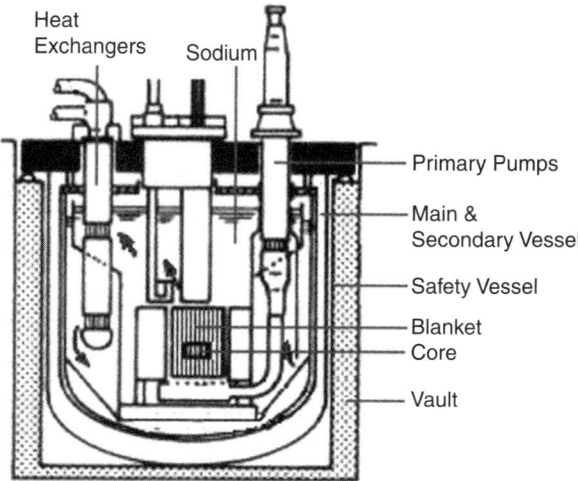

Figure 4.18. Schematic of the Phenix reactor in Marcoule, France. Adapted from WNA (2015b).

transfers the heat to a tertiary water and steam cooling loop that drives the steam turbines. The world of nuclear power generation watches this reactor (and a sister reactor under construction, the BN-800) with much interest as a part of their assessment of safety issues with fast breeder reactors and therefore with their future potential. Though there have been a number of incidents involving sodium–water interactions and a couple of sodium fires, the reactor has been repaired and resumed operation.

The Phenix was a small prototype 233 MW LMFBR constructed by the French government. Shown diagrammatically in Figure 4.18, it was a pool-type, liquid sodium–cooled reactor that began supplying electricity to the grid in 1973. This led to the construction of the larger Superphenix that began producing electricity in 1986, though it was notoriously attacked by terrorists in 1982. Despite this and other public protests, it was connected to the grid in 1994. As a result of public opposition and some technical problems, power production by the Superphenix was halted in 1996. The Phenix continued to produce power until it, too, was closed in 2009. It was the last fast breeder reactor operating in Europe.

The Clinch River Breeder Reactor was an experimental reactor designed by the U.S. government as part of an effort to examine the feasibility of the LMFBR design for commercial power generation. It was a 350 MW electric, sodium-cooled, fast breeder reactor (see Figure 4.19) whose construction was first authorized in 1970. Funding of the project was terminated in 1983, in part because of massive cost overruns. The project demonstrated the potentially high costs of constructing and operating a commercial LMFBR reactor. Moreover, in 1979, as these problems were emerging, the Three Mile Island accident (see Section 7.5.1) occurred. This clearly demonstrated that more attention needed to be paid to the safety of existing LWR plants and highlighted the potentially more serious safety issues associated with LMFBRs (see Section 7.6.3). Despite these issues, the potential technical advantages of the breeder reactor cycle mean that this design will merit further study in the years ahead.

Although virtually all present-day LMFBRs operate with uranium – plutonium oxide fuel, there is considerable interest in the future use of fuel composed of uranium – plutonium carbide, because large breeding ratios are possible with this kind of fuel. This, in turn, is because whereas there are two atoms of oxygen per atom of uranium in the oxide, there is only one atom of carbon per uranium atom in the carbide. Light atoms such as carbon and oxygen tend to moderate fission neutrons, and because there are fewer of the atoms in the carbide than in the oxide, it follows that the energy distribution of neutrons in a carbide-fueled LMFBR is shifted to energies higher than in a comparable oxide-fueled reactor.

## 4.9 Generation IV Reactors

It is appropriate to conclude this chapter with some brief description of the ideas and designs being considered in Generation IV reactors. As with the Generation II

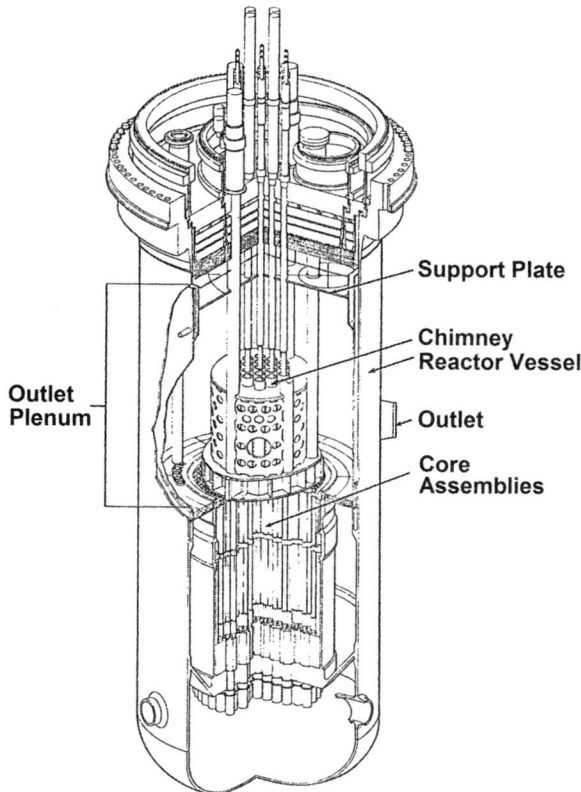

Support Plate

Chimney
Reactor Vessel

Outlet
Plenum

Outlet

Core
Assemblies

Figure 4.19. Clinch River breeder reactor. Adapted from CRBRP (1976).

reactors described earlier, Generation IV reactors are most conveniently divided into thermal reactors and fast reactors, though a number of proposed thermal reactor designs using faster neutron speeds might, more properly, be referred to as *epithermal reactors* to indicate those enhanced speeds.

### 4.9.1 Generation IV Thermal Reactors

Three Generation IV thermal reactors have received substantial attention:

- The design known as the VHTR or very high temperature reactor uses either helium or molten salt as the coolant and graphite as the moderator. Coolant temperatures at the outlet from the reactor as high as 1000°C are visualized to achieve high thermal efficiencies (hence the helium or molten salt coolant) and to allow various direct applications such as hydrogen production. Several reactor core designs have been investigated, including a prismatic block design and a pebble bed reactor design.
- The design known as the MSR or molten salt reactor is a reactor concept in which the nuclear fuel is dissolved in the molten fluoride salt coolant as uranium

tetrafluoride ($UF_4$) or thorium tetrafluoride ($ThF_4$). The fluid would then reach criticality by flowing into a graphite moderator core. The principle could be used for thermal, epithermal, or fast reactors.

- The design known as the SCWR or supercritical-water-cooled reactor is an epithermal reactor that uses supercritical water at higher pressures and temperatures as the coolant to achieve higher thermal efficiencies. It is basically an advanced LWR.

### 4.9.2  Generation IV Fast Reactors

Fast reactors have the advantage over thermal reactors that they can be configured to "burn" almost all the unstable, radioactive fission products and therefore drastically reduce the fraction of these products in the spent nuclear fuel produced by Generation II LWRs. In this way, the nuclear fuel cycle can be closed (see Section 2.2). Alternatively, they can be configured to produce more fuel than they consume. Given these potential advantages, three Generation IV fast reactors have received significant attention:

- GFRs or gas-cooled fast reactors that are helium-cooled and future evolutions of the GCR and HTGR (see Section 4.6) continue to receive substantial design attention. The typical GFR with an outlet temperature of 850°C is an evolution of the earlier mentioned VHTR to a fast neutron spectrum and a more sustainable fuel cycle.
- The sodium-cooled reactor or SFR builds on the existing LMFBR technology aiming at improving the efficiency of uranium usage and closing the nuclear fuel cycle. It is designed to use as fuel any combination of uranium, plutonium or the "nuclear" waste of LWRs.
- As an alternative to the sodium-cooled reactor and the safety problems with sodium, the lead-cooled fast reactor or LFR has received design attention. Typically the LFR would be of the pool type, with outlet temperatures of the order of 550°–800°C.

REFERENCES

Cameron, I. R. (1982). *Nuclear fission reactors*. Plenum Press.
Collier, J. G., and Hewitt, G. F. (1987). *Introduction to nuclear power*. Hemisphere.
CRBRP (1976). Clinch river breeder reactor plant technical review. CRBRP-PMC Rep. 76-06.
Duderstadt, J. J., and Hamilton, L. J. (1976). *Nuclear reactor analysis*. John Wiley.
Gregg King, C. D. (1964). *Nuclear power systems*. Macmillan.
Todres, N. E., and Kazimi, M. S. (1990). *Nuclear systems I. Thermal hydraulic fundamentals*. Hemisphere.
USNRC (1975). Reactor safety study: An assessment of accident risks in U.S. commercial nuclear power plants. *U.S. Nuclear Regulatory Commission* Rep. WASH-1400.
USAEC (1973). The safety of nuclear power reactors (light water cooled) and related facilities. *U.S. Atomic Energy Commission* Rep. WASH-1250.

Wilson, R. (1977). Physics of liquid metal fast breeder reactor safety. *Review of Modern Physics*, **49**, 893–924.

Winterton, R. H. S. (1981). *Thermal design of nuclear reactors*. Pergamon Press.

WNA. (2015a). *World Nuclear Association website*. http://www.world-nuclear.org/info/Nuclear-Fuel-Cycle/Power-Reactors/Appendices/RBMK-Reactors/.

WNA. (2015b). *World Nuclear Association website*. http://www.world-nuclear.org/info/Nuclear-Fuel-Cycle/Power-Reactors/nuclear-power-reactors/.

# 5

## Core Heat Transfer

### 5.1 Heat Production in a Nuclear Reactor

#### 5.1.1 Introduction

In this chapter, the heat transfer processes within a normally operating reactor core are analyzed and hence the conditions in the core during the normal power-production process are established. It is best to begin with an individual fuel rod and gradually move outward toward an overview of the entire core. For more detailed analyses, the reader is referred to texts such as Gregg King (1964), Tong and Weisman (1970), Todres and Kazimi (1990), and Knief (1992).

#### 5.1.2 Heat Source

As discussed earlier in Section 2.3.1, heat is produced within a nuclear reactor as a result of fission. The energy released is initially manifest primarily as the kinetic energy of fission products, of fission neutrons, and of gamma radiation. Additional energy is released as the fission products later decay, as discussed in Section 2.3.4. The kinetic energy is then converted to thermal energy as a result of the collisions of the fission products, fission neutrons, and gamma radiation with the rest of molecules in the reactor core. The majority of this energy (about 80 percent) is derived from the kinetic energy of the fission products. The fission neutrons and gamma radiation contribute about another 6 percent of the immediate heat production. This immediate energy deposition is called the *prompt heat release* to distinguish it from the subsequent, *delayed heat release* generated by the decay of the fission products. This decay heat is significant and contributes about 14 percent of the energy in an operating thermal reactor. As discussed earlier in Section 2.4.2, the fission product decay not only produces heat during normal reactor operation but that heat release continues for a time after reactor shutdown. Typically, after shutdown, the heat production decreases to 6.5 percent after 1 s, 3.3 percent after 1 min, 1.4 percent after 1 hr, 0.55 percent after 1 d, and 0.023 percent after 1 yr.

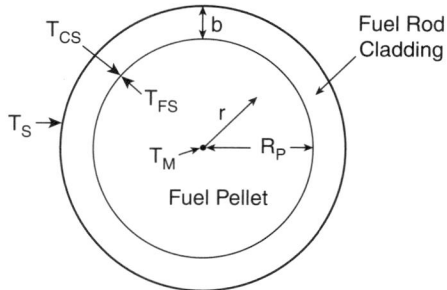

Figure 5.1. Schematic of the cross section of a fuel pellet and fuel rod.

Most of this chapter focuses on how the heat deposited in the core is transferred out of the fuel and into the core during normal reactor operation. Because almost all of the heat deposited, whether prompt or delayed, is proportional to the neutron flux, it will be assumed in the rest of this chapter that the rate of heat production is directly proportional to that neutron flux. Because the mean free path of the neutrons is large compared with the fuel rod dimensions, the neutron flux distribution is nearly uniform over the cross section of the rod though the flux in the center is somewhat less than at larger radii (because thermal neutrons that enter the fuel from the moderator or coolant are absorbed in greater number near the surface of the fuel). For present purposes, it will be assumed that flux is uniform over the cross section of the fuel rod and therefore the rate of fission and, to a first approximation, the rate of production of heat is uniform within a fuel pellet. Thus the first component of the analysis that follows concentrates on how the heat is transferred from an individual fuel rod to the surrounding coolant.

However, the neutron flux does vary substantially from one fuel rod to another within the reactor core. Consequently, the second component of the analysis that follows focuses on how the heat transfer varies from point to point within the reactor core.

### 5.1.3 Fuel Rod Heat Transfer

Consider first the heat transfer within an individual fuel rod. The cross section of a fuel pellet is sketched in Figure 5.1. The fuel pellet radius and thermal conductivity are denoted by $R_f$ and $k_f$ and the fuel rod cladding thickness and thermal conductivity by $b$ and $k_C$. The temperatures in the center of the fuel rod, at the outer surface of the fuel pellet, at the inner surface of the cladding, and at the outer surface of the fuel rod are denoted by $T_M$, $T_{FS}$, $T_{CS}$, and $T_S$, respectively. A small gap and/or a contact resistance is assumed so that $T_{FS} \neq T_{CS}$. It will also be assumed that the gradients of temperature in the axial direction are small compared with those in the radial direction and therefore that the primary heat flux takes place in the radial plane of Figure 5.1. Consequently, if the rate of heat production per unit length of a fuel rod is denoted by $Q$ and if this is uniformly distributed over the cross section of the rod, then, in steady state operation, the radially outward heat flux (per unit area) through the radial location, $r$, must be $Qr/2\pi R_f^2$. Consequently, the heat conduction equation becomes

$$\frac{Qr}{2\pi R_f^2} = -k\frac{\partial T}{\partial r} \tag{5.1}$$

where $T(r)$ is the temperature distribution and $k$ is the local thermal conductivity ($k_f$ or $k_C$). Integrating in the fuel pellet, it follows that for $0 < r < R_f$,

$$T(r) = T_M - \frac{Q}{4\pi R_f^2 k_f}r^2 \tag{5.2}$$

where the condition that $T = T_M$ at $r = 0$ has been applied. Consequently, the temperature at the surface of the fuel pellet is

$$T_{FS} = T_M - \frac{Q}{4\pi k_f} \tag{5.3}$$

As a typical numerical example, note that with a typical value of $Q$ of 500 W/cm and a thermal conductivity of $UO_2$ of $k_f = 0.03$ W/cm°K, the temperature difference between the surface and center of the fuel becomes 1400 K, a very substantial difference.

Assuming that the small gap and/or contact resistance between the fuel and the cladding gives rise to a heat transfer coefficient, $h^*$, where

$$k_f\left(\frac{\partial T}{\partial r}\right)_{r=R_f \text{ in fuel}} = k_C\left(\frac{\partial T}{\partial r}\right)_{r=R_f \text{ in cladding}} = -h^*\{T_{FS} - T_{CS}\} \tag{5.4}$$

it follows that

$$T_{CS} = T_M - \frac{Q}{4\pi k_f} - \frac{Q}{2\pi R_f h^*} \tag{5.5}$$

Integration of Equation 5.1 into the cladding ($R_f < r < R_f + b$) leads to

$$T(r) = C - \frac{Q}{4\pi R_f^2 k_C}r^2 \tag{5.6}$$

where $C$ is an integration constant. Applying the condition that $T = T_{CS}$ at $r = R_f$ yields a value for $C$ and, finally, the fuel rod surface temperature is obtained as

$$T_S = T_M - \frac{Q}{4\pi}\left[\frac{1}{k_f} + \frac{2}{h^*R_f} + \frac{\{(1 + b/R_f)^2 - 1\}}{k_C}\right] \tag{5.7}$$

Typical temperature differences in a LWR, across the fuel–cladding gap, across the cladding, and between the cladding surface and the bulk of the coolant might be of the order of 200 K, 80 K, and 15 K, respectively, so that the temperature difference between the water and the center of the fuel pellet is dominated by the temperature difference in the fuel and has a magnitude of about 1400 K. In summary, the radial temperature distribution in a fuel rod is given by Equations 5.2, 5.3, 5.5, 5.6, and 5.7, and the general form of this distribution is illustrated in Figure 5.2.

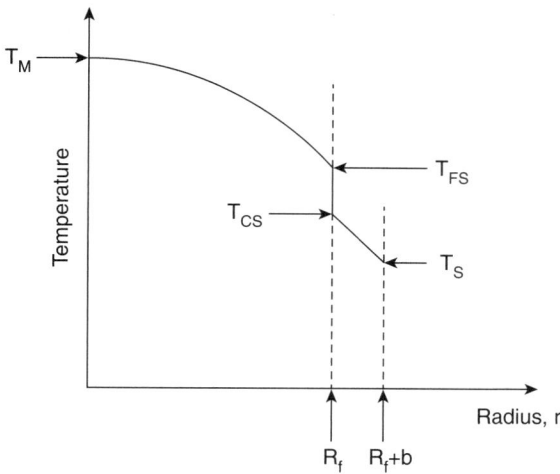

Figure 5.2. The general form of the radial temperature distribution within a fuel rod.

Because the objective is to extract heat from the fuel, it is desirable to maintain a large heat production rate, $Q$, using a proportionately large neutron flux. A large $Q$ and therefore a large power density is desirable for several reasons. First it minimizes the size of the reactor core for a given power production level and thereby reduces the cost of the core and the cost and size of the rest of the structure that contains the core. Second, higher temperature differences across the core lead to higher thermal efficiencies in the turbines driven by the coolant.

However, a high $Q$ implies large temperature differences within the fuel rods and therefore high temperatures. Thus limiting design factors are the maximum allowable temperature in the fuel, $T_M$, which must be much less than the melting temperature and, similarly, a maximum temperature in the cladding, $T_{CS}$. Moreover, the temperature of the wall in contact with the coolant, $T_S$, will also be constrained by boiling limits in the coolant. Any or all of these factors will limit the heat production because the temperature differences are proportional to $Q$. It is also clear that the temperature differences for a given heat production per unit fuel volume (or a given neutron flux) are reduced by decreasing the size of the fuel pellets, $R_f$. However, to yield the required power from the reactor, this means increasing the number of fuel rods, and this increases the cost of the core. Consequently, a compromise must be reached in which the number of fuel rods is limited but the temperature differences within each rod are maintained so as not to exceed a variety of temperature constraints.

It is valuable to list some secondary effects that must also be factored into this fuel rod analysis:

- The neutron flux in the center of the fuel rod is somewhat less than at larger radii because thermal neutrons that enter the fuel from the moderator or coolant are absorbed in greater number near the surface of the fuel. This helps even out the temperature distribution in the fuel.
- The fuel is often $UO_2$, whose manufacture causes small voids that decrease the thermal conductivity of the pellet and increase the temperature differences.

- As the fuel is used up, the gap between the fuel pellet and the cladding tends to increase, causing a decrease in $h^*$ and therefore an increase in the temperature of the fuel.
- The thermal conductivity of the fuel increases with temperature and therefore, as the heat production increases, the temperature differences in the fuel increase with $Q$ somewhat less than linearly.
- Fission gases are released by nuclear reactions in the fuel, and this can lead to significant buildup of pressure within the fuel rods that are, of course, sealed to prevent release of these gases. The gas release increases rapidly with temperature, and hence there is an important design constraint on the fuel temperature that is required to limit the maximum pressure in the fuel rods. This constraint is often more severe than the constraint that $T_M$ be less than the fuel melting temperature.

Despite these complicating factors, it is useful to emphasize that the leading constraint is the maximum allowable temperature in the center of the fuel, as is discussed in Sections 5.3 and 5.4.

### 5.1.4 Heat Transfer to the Coolant

It is appropriate at this juncture to give a brief summary of the heat transfer to the coolant to complete this review of the temperature distribution in the reactor core. In the notation of Section 5.1.3, the heat flux, $\dot{q}$, from the fuel rod to the coolant per unit surface area of the fuel rod is given by $Q/\mathcal{P}$, where $\mathcal{P}$ is the cross-sectional perimeter of the fuel rod. Though it is overly simplistic, the easiest way to relate the temperature differences in the coolant to this heat flux is by defining a heat transfer coefficient, $h$, as

$$\dot{q} = \frac{Q}{\mathcal{P}} = h(T_S - T_C) \tag{5.8}$$

where $T_S$ and $T_C$ are, respectively, the local temperature of the surface of the fuel cell and the local temperature of the coolant far from that surface. The coefficient, $h$, is, however, a complicated function of the transport properties of the coolant and of the coolant channel geometry. To express this function, a dimensionless heat transfer coefficient known as the Nusselt number, $Nu$, is introduced, defined by $hD_h/k_L$, where $D_h$ is the *hydraulic diameter* of the coolant channel (see Section 6.3.4) and $k_L$ is the thermal conductivity of the coolant. The hydraulic diameter is 4 times the cross-sectional area of the channel divided by the perimeter of that cross-sectional area and applies to a range of cross-sectional geometries of the coolant channel. The other parameters needed are the *Reynolds number*, $Re$, of the channel flow defined by $Re = \rho_L U D_h/\mu_L$, where $U$ is the volumetrically averaged coolant velocity, $\rho_L$ and $\mu_L$ are the density and viscosity of the coolant and the *Prandtl number*, $Pr$, defined by $Pr = \mu_L c_p/k_L$, where $c_p$ is the specific heat of the coolant. It transpires that $Nu$ is a function of both $Re$ and $Pr$; that functional relation changes depending on a number of factors, including whether the Prandtl number is large or small and on whether the channel flow is laminar or turbulent. Commonly used correlations are of the form $Nu = CPr^{C_1}Re^{C_2}$, where $C, C_1$, and $C_2$ are *constants*. For details of these correlations, the reader is referred to heat transfer texts (e.g., Rohsenow and Hartnett 1973). For simplicity and illustrative purposes, it will be assumed that $h$ is a known constant that, in the absence of boiling, is uniform throughout the reactor core. The case of boiling, either in a boiling water reactor or during an excursion in a normally nonboiling reactor, is covered in a later section.

The next step is to subdivide the coolant flow through the reactor core into a volume flow rate, $\dot{V}$, associated with each individual fuel rod. As that flow proceeds through the core, it receives heat from the fuel rod at a rate of $Qdz$ for an elemental length, $dz$, of the rod. As a result, the temperature rise in the coolant over that length is $dT$, where

$$\rho_L \dot{V} c_p \frac{dT}{dz} = Q \tag{5.9}$$

To obtain the temperature distribution over the length of a coolant channel, it is necessary to integrate the relation 5.9. To do so, the variation of $Q$ with $z$ is needed. This is roughly proportional to the variation of the neutron flux with $z$. As seen in Chapter 3, the neutron flux distribution also varies with the radial location, $r$, within the reactor core; it also depends on control factors such as the extent of the control rod insertion (Section 3.7.4).

## 5.2 Core Temperature Distributions

As a representative numerical example of the temperature distribution in a reactor core, consider a homogeneous cylindrical reactor without reflectors and without control rod insertion. The neutron flux has the form given by Equation 3.41 (coordinates defined in Figure 3.1), and therefore $Q$ will be given by

$$Q = Q_M \cos\left(\frac{\pi z}{H_E}\right) J_0\left(\frac{2.405r}{R_E}\right) \tag{5.10}$$

where the constant $Q_M$ is the maximum value at the center of the reactor core. Note for future use that the average heat flux would then be about $0.4Q_M$. Substituting Equation 5.10 into Equation 5.9 and integrating, the temperature of the coolant, $T_C$, within the reactor core becomes

$$T_C = T_{CI} + \frac{Q_M H_E}{\pi \rho_L \dot{V} c_p} J_0\left(\frac{2.405r}{R_E}\right)\left[\sin\left(\frac{\pi z}{H_E}\right) + 1\right] \tag{5.11}$$

where $T_{CI}$ is the coolant inlet temperature. The form of this temperature distribution along the centerline of the reactor core ($r = 0$) is shown labeled $\beta = 0$ in Figure 5.3. Similar integrations can readily be performed for the neutron flux distributions at various control rod insertions (see Section 3.7.4), and three such examples are also included in Figure 5.3. Note that the temperature rise in the upper part of the core is reduced due to the decrease in the heat production in that part of the reactor.

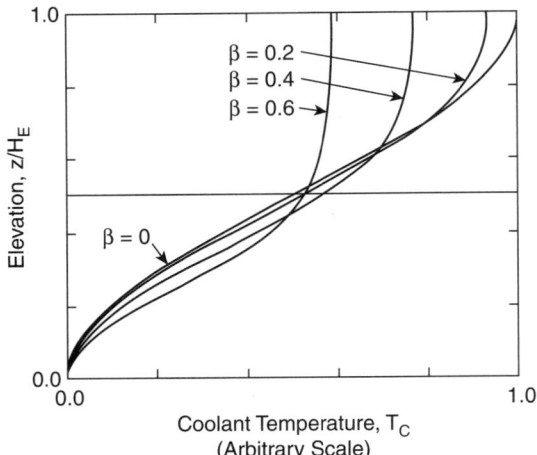

Figure 5.3. Axial coolant temperature distributions within a cylindrical reactor where the horizontal scale may differ for each line plotted. Solid lines: $\beta = 0$ is a homogeneous reactor and the lines for $\beta = 0.2, 0.4$, and $0.6$ are for various control rod insertions corresponding to the neutron fluxes in Figure 3.7 (for the case of $H_E/R_E = 2.0$ and $L_2/R_E = 0.36$).

It is important to emphasize that, even in the absence of boiling (addressed in Section 5.5), these calculations of the axial temperature distribution are only of very limited validity. In practical reactors, variations in the fuel and moderator distributions are used to even out the heat distribution. Moreover, thermal and transport properties like the heat transfer coefficient may vary significantly within the reactor core. Nevertheless the preceding calculations combined with the knowledge of the radial distribution of temperature implicit in Equation 5.11 and coupled with the temperature distribution within each fuel rod as described in Section 5.1.3 allow construction of the temperatures throughout the core in a way that is qualitatively correct.

## 5.3 Core Design: An Illustrative LWR Example

The results of the last few sections allow presentation of a simplistic but illustrative design methodology for the reactor core. In this section, an illustrative LWR example is examined; the next section presents an LMFBR example.

For the sake of this simplified numerical evaluation of a LWR core, it is stipulated that the maximum temperature in the fuel must be well below the melting temperature of uranium dioxide, specifically much less than about 3000 K. Consequently the usual red-line design maximum is in the range 2000–2300 K. Because the maximum coolant temperature is about 500 K and the maximum temperature difference between the center of the fuel rod and the coolant is therefore about 1500–1800 K, this effectively limits the heat flux from the fuel rod for a given radius, $R_f$, of that rod. From this perspective, the smaller the rod the greater the potential power output but there are other considerations (such as the structural strength and the neutronics) that necessarily limit how small the fuel rod radius can be. These compromises led to fuel rod radii, $R_f$, of 0.53 cm and 0.71 cm, respectively, for the typical PWR and BWR.

Then Equation 5.7 (or 5.3) determines the maximum heat flux allowable in the reactor. For a fuel thermal conductivity of $k_f = 0.03$ W/cm° K, these equations yield a maximum allowable value of $Q$ of about 430 W/cm in the hottest part of the core. This, in turn, implies a red line value for the *average* heat flux of about 180 W/cm.

The next step is to stipulate the desired ratio of moderator volume to fuel volume, $\alpha_{mf}$. This is primarily determined by nuclear considerations that dictate a moderator to fuel volume ratio of $\alpha_{mf} \approx 1$.

The objective in this example will be to find the size of the cylindrical reactor needed for a 1150 MW electric power plant with efficiency of 34 percent so that the thermal power generated by the core is $P = 3400$ MW. The target is a cylindrical reactor of diameter, $2R$, and a height equal to that diameter. Then the fuel volume (neglecting the cladding volume) will be $2\pi R^3/(1 + \alpha_{mf})$ and the required number of fuel rods, $N_f$, of the same height as the reactor will be

$$N_f = R^2/[R_f^2(1 + \alpha_{mf})] \tag{5.12}$$

Moreover, the thermal power of the reactor power, $P$, will clearly be given by the heat added to the coolant during its passage through the core or

$$P = 2RQ_{av}N_f \tag{5.13}$$

where $Q_{av}$ is the average heat flux per unit fuel rod length, averaged over the volume of the reactor. If the maximum value of that heat flux is set at 420 W/cm (see earlier), then a reasonable, illustrative value of this average would be $Q_{av} = 180$ W/cm. Substituting this value into Equation 5.13 as well as the expression 5.12 for $N_f$ and $P = 3400$ MW yields an expression for the dimension of the reactor, $R$. For the aforementioned values of $\alpha_{mf}$ and $R_f$, this yields

$$\text{For PWR:} \quad R = 1.7 \text{ m} \quad \text{For BWR:} \quad R = 2.0 \text{ m} \tag{5.14}$$

values that are close to the actual volumetric-equivalent radii of 1.7 m and 1.8 m for the typical PWR and BWR, respectively. Despite the crudeness of these calculations, they come close to the dimensions of light water reactor cores.

In addition, substitution back into Equation 5.12 yields $N_f \approx 54,000$ and $N_f \approx 46,000$ for the PWR and the BWR, respectively, values that are again close to the actual typical numbers of fuel rods, namely, 56,000 and 47,000, respectively.

## 5.4 Core Design: An LMFBR Example

An illustrative LMFBR core design follows very similar lines though with numerical differences. The chosen fuel rod diameters are significantly smaller to allow higher heat fluxes (typical fuel rod radii are 0.38 cm). Liquid sodium coolant temperatures of the order of 820° K mean a maximum temperature difference between the fuel rod center and the sodium coolant of about 1500° K. According to Equation 5.3,

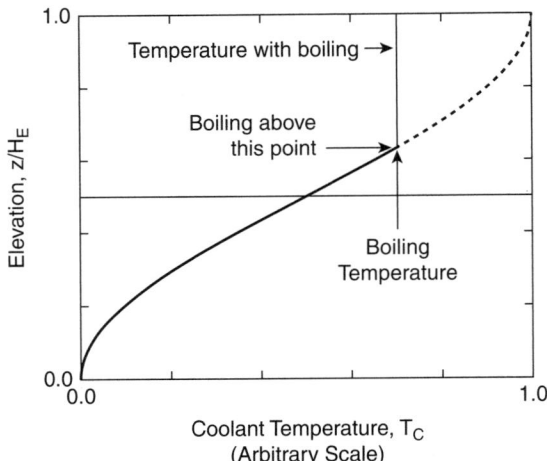

Figure 5.4. Typical modification of the axial coolant temperature distribution due to boiling, where the curve below the boiling point is reproduced from Figure 5.3.

this implies a maximum heat flux of 490 W/cm. If an *average* maximum heat flux of 290 W/cm is chosen, Equations 5.13 and 5.12 imply a reactor radius $R$ of 1.1 m for a 2600 MW thermal generating plant (it has again been assumed that the fuel takes up one-half of the volume of the core). This reactor radius is close to the actual, typical volumetric-equivalent radius of an LMFBR core of 1.1 m, much smaller than a LWR core of the same power.

## 5.5 Boiling Water Reactor

### 5.5.1 Temperature Distribution

If the temperature of the coolant reaches the boiling point before the top of the reactor, then virtually all the heat generated will go into latent heat to produce vapor, and the temperature above that boiling point elevation will remain approximately constant, as illustrated in Figure 5.4 (an adaption of Figure 5.3). This is because the pressure change is small, and so the thermodynamic state of the multiphase fluid remains at approximately the same saturated temperature and pressure while the mass quality of the *steam* flow, $\mathcal{X}$, increases with elevation (the mass quality, $\mathcal{X}$, is defined as the ratio of the mass flux of vapor to the total mass flux; see Section 6.2.1). This relative constancy of the pressure and temperature will hold until all the liquid has evaporated. Of course, if the critical heat flux is reached (see Sections 6.5.2 to 6.5.4 and 5.6) and film boiling (see Sections 6.5.5 and 6.5.6) sets in, the fuel rod temperature will rise rapidly and the potential for meltdown could exist. This critical accident scenario is discussed in Chapter 7.

Above the elevation at which boiling starts, and assuming that the critical heat flux is not reached, it is roughly true that all the heat flux from the fuel rods, $\mathcal{Q}$, is converted to latent heat. Therefore it follows that the rate of increase of the mass quality, $d\mathcal{X}/dz$, in the coolant flow will be given by

$$\frac{d\mathcal{X}}{dz} = \frac{\mathcal{Q}}{\dot{m}\mathcal{L}}$$ (5.15)

where $\dot{m}$ is the mass flow rate per fuel rod (equal to $\rho_L \dot{V}$ below the boiling elevation) and $\mathcal{L}$ is the latent heat of the coolant. Because the temperature and pressure do not change greatly above the boiling point elevation, the latent heat, $\mathcal{L}$, is also relatively constant, and therefore Equation 5.15 can be written in the integrated form

$$\mathcal{X} = \frac{1}{\dot{m}\mathcal{L}} \int_{z_B}^{z} \mathcal{Q}dz$$ (5.16)

where $z_B$ is the elevation at which boiling starts and where the mass quality is therefore zero. Note that the rate of increase of the mass quality decreases with the mass flow rate, $\dot{m}$, and increases with the heat flux, $\mathcal{Q}$.

The evaluation of the mass quality (and other multiphase flow properties) is important for a number of reasons, all of which introduce a new level of complexity to the analysis of the core neutronics and thermo-hydraulics. In the next section, consideration is given to how the calculations of these quantities might proceed.

### 5.5.2 Mass Quality and Void Fraction Distribution

Boiling in the flow channels changes the moderating properties of the fluid and hence the reactivity, and this, in turn, will change the heat flux. Consequently, it is necessary to perform simultaneous neutronics and multiphase flow calculations to properly establish the heat flux and two-phase flow conditions in the boiling region. Perhaps it is most illustrative to consider approaching the solution iteratively starting with the heat flux distribution that would occur in the absence of boiling (see Section 5.2), as sketched with the solid line in the top graph of Figure 5.5. This would imply a coolant temperature given by the solid line to the left of the boiling location in the second graph. It will be assumed that when this reaches the saturated vapor temperature at the prevailing coolant pressure, boiling begins and the temperature thereafter remains at the saturated vapor temperature (because the pressure decreases with elevation due to a combination of hydrostatic pressure drop and frictional pressure drop, the saturated vapor temperature may drop a little, as sketched in Figure 5.5). For the present it will be assumed that the critical heat flux (CHF) (see Section 6.5.2) is not reached in the reactor core; otherwise, the temperature would begin to rise substantially, as sketched by the dashed line in the second graph of Figure 5.5.

The next step is to integrate the heat flux using Equation 5.13 to obtain the mass quality as a function of elevation, as sketched in the third graph of Figure 5.5; note that the mass quality, $\mathcal{X}$, will begin at zero at the point where boiling begins and that the slope of the line beyond that point will vary like the heat flux, $\mathcal{Q}$. The next step is to deduce the void fraction, $\alpha$, of the two-phase flow knowing the mass quality, $\mathcal{X}$. This is a more complex step for, as discussed in Section 6.2.1, the relation between $\alpha$ and $\mathcal{X}$ involves the velocities of the two phases, and these may be quite different.

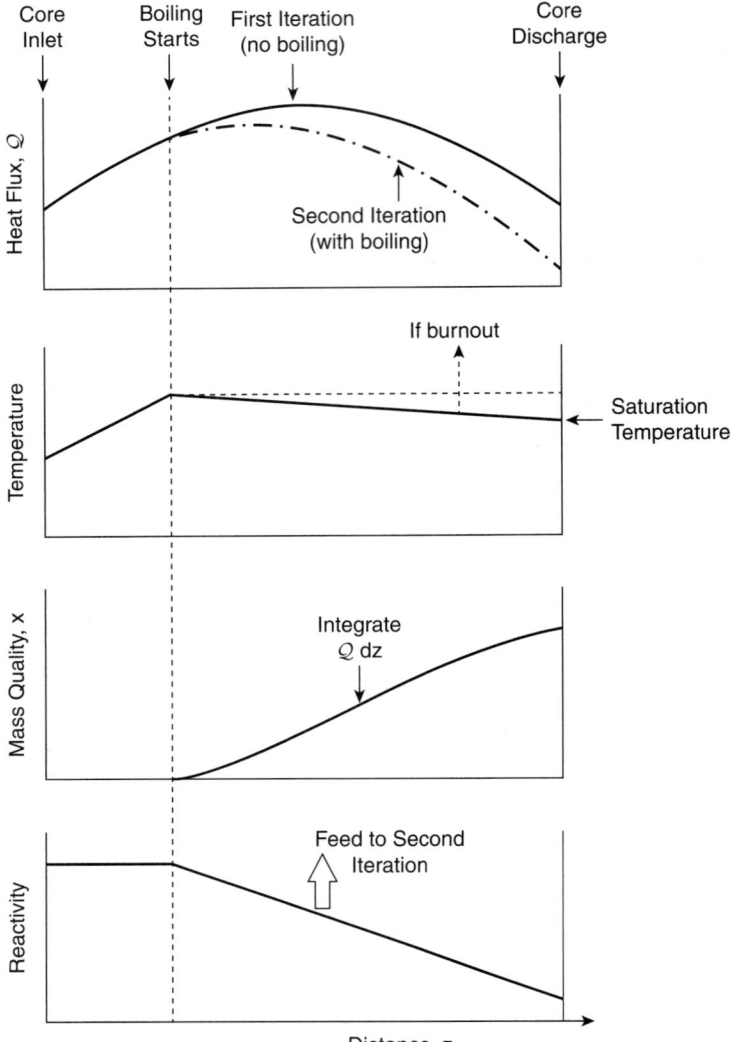

Figure 5.5. Schematic relation between the heat flux, $Q$, as a function of elevation within the core of a boiling water reactor (top) and the coolant temperature, mass quality, and reactivity.

The calculation of the void fraction is necessary because the void fraction changes the moderating properties of the two-phase coolant. The local reactivity will decline as $\alpha$ increases, as discussed in Section 7.1.2, and will therefore take the qualitative form sketched in the lowest graph of Figure 5.5.

However, this change in the reactivity means that the heat flux will be different from that which was assumed at the start of the calculation. Therefore the second iteration needs to begin with a revised heat flux determined using the new, corrected reactivity. This will result in a decreased heat flux above the location of boiling initiation, and the previous series of steps then need to be repeated multiple times until a converged state is reached.

It should be noted that the two-phase flow also alters the heat transfer coefficient, $h$, governing the heat flux from the fuel rods to the coolant. Under these conditions, the functional relation between the Nusselt number, $Nu$, and the Reynolds and Prandtl numbers will change, and this, in turn, will change the temperatures in the fuel rod. This complication also needs to be factored into the preceding calculation.

## 5.6 Critical Heat Flux

In the preceding two sections, it was assumed that the critical heat flux conditions and temperatures were not reached within the reactor. Indeed, care is taken to stay well below those temperatures during the designed normal operation of a boiling water reactor. However, because a postulated accident in a PWR, a BWR, or any other liquid-cooled reactor core might result in enhanced boiling, analyses similar to those described in the preceding sections need to be carried out to predict the evolution of that accident scenario. If burnout and critical heat flux conditions were to occur at some elevation within the core, this would further modify the conditions described in the last section. The coolant and fuel rod temperatures above that burnout location would rise rapidly, as would the mass quality of the coolant, which would approach unity. However, this would result in yet another decrease in the reactivity and therefore in the local heat generation within the fuel. Moreover, in a loss-of-coolant accident or LOCA (see Section 7.3), an evolving decrease in the coolant flow rate, $\dot{m}$, will result in an enhanced rate of increase in the mass quality (as illustrated in Equation 5.13), and this would promote the chance of burnout.

Because of the potential for fuel rod damage and meltdown in such a postulated accident scenario, it is very important to be able to predict the evolution of such an event. The preceding description of how such a calculation might proceed only serves to indicate what a complicated multiphase flow calculation that involves. Further comment on these efforts is included at the end of the next chapter.

### REFERENCES

Gregg King, C. D. (1964). *Nuclear power systems.* Macmillan.
Knief, R. A. (1992). *Nuclear engineering: Theory and practice of commercial nuclear power.* Hemisphere.
Rohsenow, W. M., and Hartnett, J. P. (1973.) *Handbook of heat transfer.* McGraw-Hill.
Todres, N. E., and Kazimi, M. S. (1990). *Nuclear systems I: Thermal hydraulic fundamentals.* Hemisphere.
Tong, L. S., and Weisman, J. (1970). *Thermal analysis of pressurized water reactors.* American Nuclear Society.

# 6

# Multiphase Flow

## 6.1 Introduction

A multiphase flow is the flow of a mixture of phases or components. Such flows occur in the context of nuclear power generation either because the reactor (such as a BWR) is designed to function with a cooling system in which the primary coolant consists of several phases or components during normal operation or because such flows might occur during a reactor accident. In the latter context, predictions of how the accident might progress or how it might be ameliorated may involve analyses of complicated and rapidly changing multiphase flows. Consequently, some familiarity with the dynamics of multiphase flows is essential to the nuclear reactor designer and operator. This chapter provides a summary of the fundamentals of the dynamics of multiphase flows. In general, this is a subject of vast scope that ranges far beyond the limits of this book. Consequently, the reader will often be referred to other texts for more detailed analyses and methodologies.

## 6.2 Multiphase Flow Regimes

From a practical engineering point of view, one of the major design difficulties in dealing with multiphase flow is that the mass, momentum, and energy transfer rates and processes can be quite sensitive to the geometric distribution or topology of the components within the flow. For example, the topology may strongly affect the interfacial area available for mass, momentum, or energy exchange between the phases. Moreover, the flow within each phase or component will clearly depend on that geometric distribution. Consequently, there is a complicated two-way coupling between the flow in each of the phases or components and the geometry of the flow (as well as the rates of change of that geometry). The complexity of this two-way coupling presents a major challenge in the analysis and prediction of multiphase flows.

### 6.2.1 Multiphase Flow Notation

The notation that will be used for multiphase flow is as follows. Uppercase subscripts will refer to the property of a specific phase or component, for example, $C$ for a

90

continuous phase, $D$ for a disperse phase, $L$ for liquid, $G$ for gas, $V$ for vapor. In some contexts, generic subscripts $N, A$, or $B$ will be used for generality. Specific properties frequently used are as follows. The densities of individual components or phases are denoted by $\rho_N$. *Volumetric fluxes* (volume flow per unit area) of individual components will be denoted by $j_N$, and the *total volumetric flux* is denoted by $j = j_A + j_B$. *Mass fluxes* will then be given by $\rho_N j_N$, and velocities of the individual components or phases will be denoted by $u_N$.

The volume fraction of a component or phase is denoted by $\alpha_N$, and in the case of two components or phases, $A$ and $B$, it follows that $\alpha_B = 1 - \alpha_A$. Then the mixture density, denoted by $\rho$, is given by

$$\rho = \alpha_A \rho_A + \alpha_B \rho_B \tag{6.1}$$

It also follows that the volume flux of a component, $N$, and its velocity are related by $j_N = \alpha_N u_N$.

Two other fractional properties are the *volume quality*, $\beta_N$, defined as the ratio of the volumetric flux of the component, $N$, to the total volumetric flux so that, for example, $\beta_A = j_A / j$. Note that, in general, $\beta$ is not necessarily equal to $\alpha$. The *mass fraction*, $x_A$, of a phase or component, $A$, is simply given by $\rho_A \alpha_A / (\rho_A \alpha_A + \rho_B \alpha_B)$. Conversely, the *mass quality*, $\mathcal{X}_A$, often referred to simply as *the quality*, is the ratio of the mass flux of component, $A$, to the total mass flux, or

$$\mathcal{X}_A = \frac{\rho_A j_A}{\rho_B j_B + \rho_A j_A} \tag{6.2}$$

Furthermore, when only two components or phases are present, it is often redundant to use subscripts on the volume fraction and the qualities since $\alpha_A = 1 - \alpha_B, \beta_A = 1 - \beta_B$, and $\mathcal{X}_A = 1 - \mathcal{X}_B$. Thus unsubscripted quantities $\alpha, \beta$, and $\mathcal{X}$ will often be used in these circumstances.

Finally, note for future use that the relation between the volume fraction, $\alpha_A$, and the mass quality, $\mathcal{X}_A$, for a given phase or component, $A$, in a two-phase or two-component mixture of $A$ and $B$ follows from Equation 6.2, namely,

$$\mathcal{X}_A = \frac{\rho_A \alpha_A u_A}{\rho_B (1 - \alpha_A) u_B + \rho_A \alpha_A u_A} \tag{6.3}$$

where $u_A$ and $u_B$ are the velocities of the two phases or components. Therefore $\mathcal{X}_A$ and $\alpha_A$ may be quite different.

### 6.2.2 Multiphase Flow Patterns

An appropriate starting point in any analysis of multiphase flow is a phenomenological description of the geometric distributions that are observed in these flows. A particular type of geometric distribution of the components is called a *flow pattern* or *flow regime*, and many of the names given to these flow patterns (such as annular flow or bubbly flow) are now quite standard. Usually the flow patterns are recognized by visual inspection, though other means, such as analysis of the spectral content of the unsteady pressures or the fluctuations in the volume fraction, have been devised for those circumstances in which visual information is difficult to obtain (Jones and Zuber, 1974).

For some of the simpler flows, such as those in vertical or horizontal conduits, a substantial number of investigations have been conducted to determine the dependence of the flow pattern on component volume fluxes, $(j_A, j_B)$, on volume fraction, and on the fluid properties, such as density, viscosity, and surface tension. The results are often displayed in the form of a *flow regime map* that identifies the flow patterns occurring in various parts of a parameter space defined by the component flow rates. The flow rates used may be the volume fluxes, mass fluxes, momentum fluxes, or other similar quantities, depending on the author. Summaries of these flow pattern studies and the various empirical laws extracted from them are a common feature in reviews of multiphase flow (see, e.g., Brennen 2005; Wallis 1969; Weisman 1983).

The boundaries between the various flow patterns in a flow pattern map occur because a regime becomes unstable as the boundary is approached and growth of this instability causes transition to another flow pattern. Like the laminar-to-turbulent transition in single-phase flow, these multiphase transitions can be rather unpredictable because they may depend on otherwise minor features of the flow, such as the roughness of the walls or the entrance conditions. Hence the flow pattern boundaries are not distinctive lines but more poorly defined transition zones.

However, there are other serious difficulties with most of the existing literature on flow pattern maps. One of the basic fluid mechanical problems is that these maps are often dimensional and therefore apply only to the specific pipe sizes and fluids employed by the investigator. A number of investigators (e.g., Baker 1953; Schicht 1969; Weisman and Kang 1981) have attempted to find generalized coordinates that would allow the map to cover different fluids and conduits of different sizes. However, such generalizations can only have limited value because several transitions are represented in most flow pattern maps and the corresponding instabilities are governed by different sets of fluid properties. Hence, even for the simplest duct geometries, no universal, dimensionless flow pattern maps exist that incorporate the full, parametric dependence of the boundaries on the fluid characteristics.

Beyond these difficulties are a number of other troublesome questions. In single-phase flow, it is well established that an entrance length of 30–50 diameters is necessary to establish fully developed turbulent pipe flow. The corresponding entrance lengths for multiphase flow patterns are less well established, and it is quite possible that some of the reported experimental observations are for temporary or developing flow patterns. Moreover, the implicit assumption is often made that there exists a unique flow pattern for given fluids with given flow rates. It is by no means certain that this is the case. Consequently, there may be several possible flow patterns whose occurrence may depend on the initial conditions, specifically on the manner in which the multiphase flow is generated.

### 6.2.3 Flow Regime Maps

Despite the issues and reservations discussed in the preceding section, it is useful to provide some examples of flow regime maps along with the definitions that help

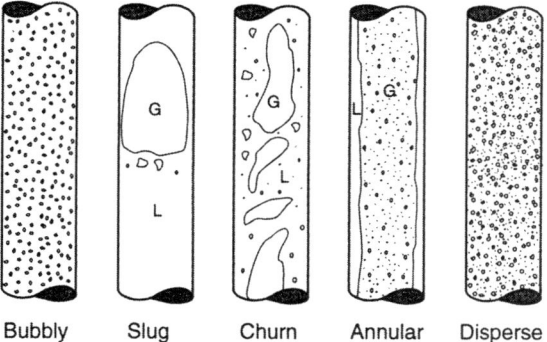

Bubbly    Slug    Churn    Annular    Disperse

Figure 6.1. Sketches of flow regimes for two-phase flow in a vertical pipe. Adapted from Weisman (1983).

distinguish the various regimes. Perhaps the most widely studied multiphase flow is that of a gas–liquid mixture in a horizontal conduit, and here some progress has been made in understanding the scaling of the boundaries in a flow regime map (see, e.g., Hubbard and Dukler 1966; Weisman 1983; Mandhane et al. 1974; Brennen 2005). However, the focus in nuclear power generation is more frequently on vertical gas–liquid flow and the typical definitions of these flow regimes are as displayed graphically in Figure 6.1 (see, e.g., Hewitt and Hall-Taylor 1970; Butterworth and Hewitt 1977; Hewitt 1982; Whalley 1987). An example of a vertical flow regime map is shown in Figure 6.2, this one using momentum flux axes rather than volumetric or mass fluxes. Note the wide range of flow rates in this flow regime map by Hewitt and Roberts (1969) and the fact that they correlated both air–water data at atmospheric pressure and steam–water flow at high pressure.

It should be added that flow regime information such as that presented in Figure 6.2 appears to be valid both for flows that are not evolving with axial distance along the pipe and for flows, such as those in a reactor, in which the volume fraction is increasing with axial position. Figure 6.3 provides a sketch of the kind of evolution one might expect in a vertical fluid passage within a reactor core based on the flow regime maps given previously.

### 6.2.4 Flow Pattern Classifications

One of the most fundamental characteristics of a multiphase flow pattern is the extent to which it involves global separation of the phases or components. At the two ends of the spectrum of separation characteristics are those flow patterns that are termed *disperse* and those that are termed *separated*. A *disperse* flow pattern is one in which one phase or component is widely distributed as drops, bubbles, or particles in the other *continuous* phase. Conversely, a *separated* flow consists of separate, parallel streams of the two (or more) phases. Even within each of these limiting states, there are various degrees of component separation. The asymptotic limit of a disperse flow in which the disperse phase is distributed as an infinite number of infinitesimally small bubbles or drops is termed a *homogeneous* multiphase flow. Because the relative velocity of a tiny bubble or drop approaches zero as its size decreases, this limit

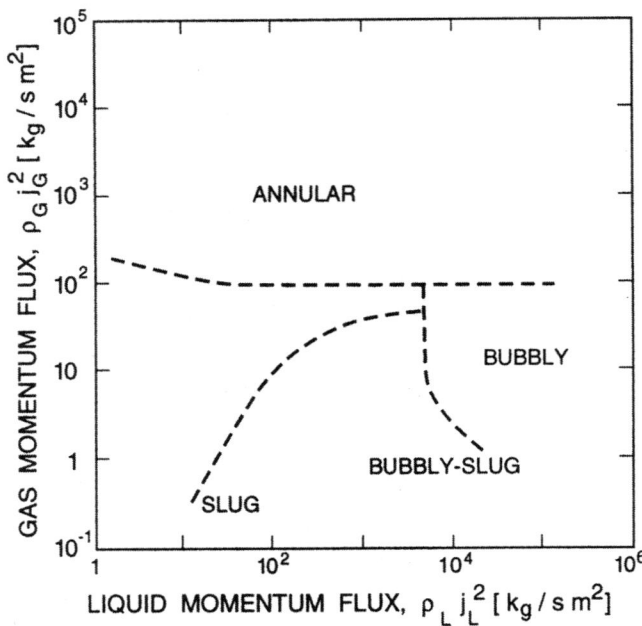

Figure 6.2. The vertical flow regime map of Hewitt and Roberts (1969) for flow in a 3.2-cm-diameter tube, validated for both air–water flow at atmospheric pressure and steam–water flow at high pressure.

implies zero relative motion between the phases. However, there are many practical disperse flows, such as bubbly or mist flow in a pipe, in which the flow is quite disperse in that the particle size is much smaller than the pipe dimensions but in which the relative motion between the phases is significant.

Within separated flows are similar gradations or degrees of phase separation. The low-velocity flow of gas and liquid in a pipe that consists of two single-phase streams can be designated a *fully separated* flow. Conversely, most annular flows in a vertical pipe consist of a film of liquid on the walls and a central core of gas that contains a significant number of liquid droplets. These droplets are an important feature of annular flow, and therefore the flow can only be regarded as partially separated.

To summarize, one of the basic characteristics of a flow pattern is the degree of separation of the phases into streamtubes of different concentrations. The degree of separation will, in turn, be determined by (1) some balance between the fluid mechanical processes enhancing dispersion and those causing segregation, or (2) the initial conditions or mechanism of generation of the multiphase flow, or (3) some mix of both effects.

A second basic characteristic that is useful in classifying flow patterns is the level of intermittency in the volume fraction. An example of an intermittent flow pattern is slug flow in a vertical pipe. The first separation characteristic was the degree of separation of the phases between streamtubes; this second intermittency characteristic can be viewed as the degree of periodic separation in the streamwise direction. The slugs or waves are kinematic or concentration waves (sometimes called continuity waves), and the reader is referred to Brennen (2005) for a general discussion of the

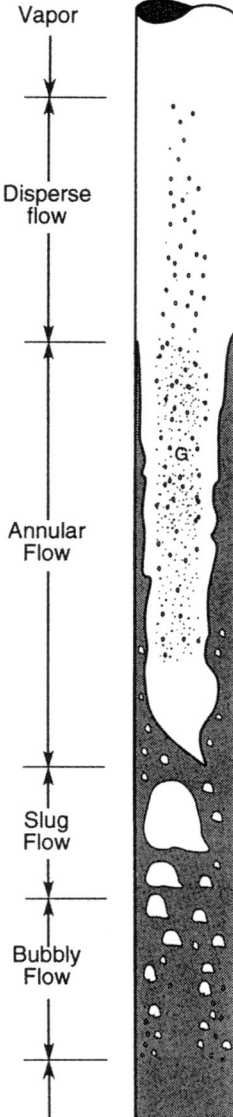

Figure 6.3. The evolution of the steam–water flow in a boiling vertical conduit.

structure and characteristics of such waves. Intermittency is the result of an instability in which kinematic waves grow in an otherwise nominally steady flow to create significant streamwise separation of the phases.

The sections that follow include a brief description of how these ideas of cross-streamline separation and intermittency can lead to an understanding of the limits of specific multiphase flow regimes. Both the limits on disperse flow regimes and the limits on separated flow regimes are briefly addressed.

### 6.2.5 Limits of Disperse Flow Regimes

To determine the limits of a disperse phase flow regime, it is necessary to identify the dominant processes enhancing separation and those causing dispersion. By far the most common process causing phase separation is due to the difference in the densities of the phases, and the mechanisms are therefore functions of the ratio of the density of the disperse phase to that of the continuous phase. Then the buoyancy forces caused either by gravity or, in a nonuniform or turbulent flow, by the Lagrangian fluid accelerations will create a relative velocity between the phases that may lead to phase separation.

Whereas the primary mechanism of phase separation in a quiescent multiphase mixture is sedimentation, in flowing mixtures, the mechanisms are more complex and, in most applications, are controlled by a balance between the buoyancy–gravity forces and the hydrodynamic forces. In high–Reynolds number, turbulent flows, the turbulence can cause either dispersion or segregation. Segregation may occur when, for example, solid particles suspended in a gas flow are centrifuged out of the more intense turbulent eddies and collect in the shear zones in between (see, e.g., Squires and Eaton 1990; Elghobashi and Truesdell 1993) or when bubbles in a liquid collect in regions of low pressure, such as in the wake of a body or in the centers of vortices (see, e.g., Pan and Banerjee 1997). Counteracting these separation processes are dispersion processes, and in many engineering contexts, the principal dispersion is caused by the turbulent or other unsteady motions in the continuous phase. The shear created by unsteady velocities can also cause either fission or fusion of the disperse phase bubbles or drops. Quantitative evaluation of these competing forces of segregation and dispersion can lead to criteria determining the boundary between separated and disperse flow in a flow regime map (see, e.g., Brennen 2005).

As a postscript, note from the preceding that an evaluation of the disperse flow separation process will normally require knowledge of the bubble or droplet size, and this is not usually known a priori. This is a serious complication because the size of the bubbles or drops is often determined by the flow itself because the flow shear tends to cause fission of those bubbles or drops and therefore limit the maximum size of the surviving bubbles or drops. Then the flow regime may depend on the particle size that in turn depends on the flow, and this two-way interaction can be difficult to unravel. When the bubbles or drops are very small, a variety of forces may play a role in determining the effective size. However, often the bubbles or drops are sufficiently large that the dominant force resisting fission is due to surface tension, whereas the dominant force promoting fission is the shear in the flow. Typical regions of high shear occur in boundary layers, in vortices, or in turbulence. Frequently, the larger drops or bubbles are fissioned when they encounter regions of high shear and do not subsequently coalesce to any significant degree. For further analyses and criteria, the reader is referred to Mandhane et al. (1974), Taitel and Dukler (1976), and Brennen (2005).

### 6.2.6 Limits on Separated Flow

Attention will now be turned to the limits on separated flow regimes, and the primary mechanism that determines that limit is the Kelvin–Helmholtz instability. Separated flow regimes such as stratified horizontal flow or vertical annular flow can

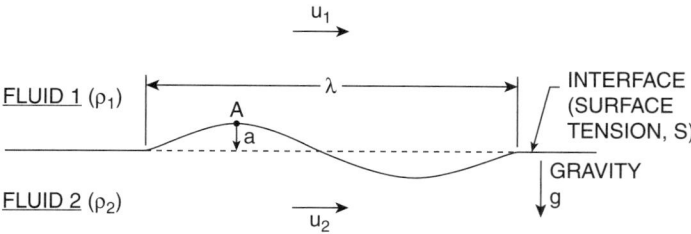

Figure 6.4. Sketch showing the notation for Kelvin–Helmholtz instability.

become unstable when waves form on the interface between the two fluid streams (subscripts 1 and 2). As indicated in Figure 6.4, the densities of the fluids will be denoted by $\rho_1$ and $\rho_2$ and the velocities by $u_1$ and $u_2$. If these waves continue to grow in amplitude, they cause a transition to another flow regime, typically one with greater intermittency and involving plugs or slugs. Therefore, to determine this particular boundary of the separated flow regime, it is necessary to investigate the potential growth of the interfacial waves, whose wavelength will be denoted by $\lambda$ (wave number, $\kappa = 2\pi/\lambda$). Studies of such waves have a long history, originating with the work of Kelvin and Helmholtz, and the phenomena they revealed have come to be called Kelvin–Helmholtz instabilities (see, e.g., Yih 1965). In general, this class of instabilities involves the interplay between at least two of the following three types of forces:

- A buoyancy force due to gravity and proportional to the difference in the densities of the two fluids. In a horizontal flow in which the upper fluid is lighter than the lower fluid, this force is stabilizing. When the reverse is true, the buoyancy force is destabilizing, and this causes Rayleigh-Taylor instabilities. When the streams are vertical, as in vertical annular flow, the role played by the buoyancy force is less clear.
- A surface tension force that is always stabilizing.
- A Bernoulli effect that implies a change in the pressure acting on the interface caused by a change in velocity resulting from the displacement, $a$, of that surface. For example, if the upward displacement of the point A in Figure 6.5 were to cause an increase in the local velocity of fluid 1 and a decrease in the local velocity of fluid 2, this would imply an induced pressure difference at the point A that would increase the amplitude of the distortion, $a$.

The interplay between these forces is most readily illustrated by a simple example. Neglecting viscous effects, one can readily construct the planar, incompressible potential flow solution for two semi-infinite horizontal streams separated by a plane horizontal interface (as in Figure 6.4) on which small amplitude waves have formed. Then it is readily shown (Lamb1879; Yih 1965) that Kelvin–Helmholtz instability will occur when

$$\frac{g\Delta\rho}{\kappa} + \mathcal{S}\kappa - \frac{\rho_1\rho_2(\Delta u)^2}{\rho_1 + \rho_2} < 0 \qquad (6.4)$$

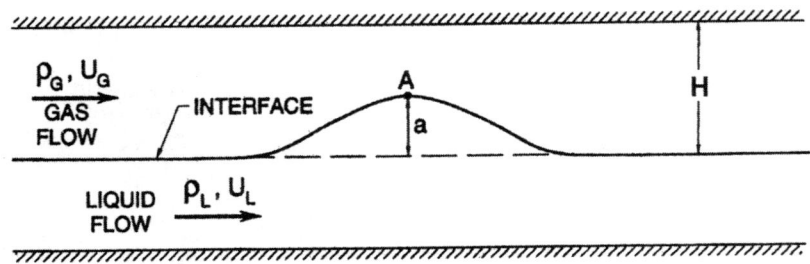

Figure 6.5. Sketch showing the notation for stratified flow instability.

where $S$ is the surface tension of the interface. The contributions from the three previously mentioned forces are self-evident. Note that the surface tension effect is stabilizing because that term is always positive, the buoyancy effect may be stabilizing or destabilizing depending on the sign of $\Delta\rho$, and the Bernoulli effect is always destabilizing. Clearly one subset of this class of Kelvin–Helmholtz instabilities are the Rayleigh–Taylor instabilities that occur in the absence of flow ($\Delta u = 0$) when $\Delta\rho$ is negative. In that static case, the preceding relation shows that the interface is unstable to all wave numbers less than the critical value, $\kappa = \kappa_c$, where

$$\kappa_c = \left(\frac{g(-\Delta\rho)}{S}\right)^{\frac{1}{2}} \tag{6.5}$$

The Bernoulli effect is frequently the primary cause of instability in a separated flow and can lead to transition to a plug or slug flow regime. As a first example of the instability induced by the Bernoulli effect, consider the stability of the horizontal stratified flow depicted in Figure 6.5, where the destabilizing Bernoulli effect is primarily opposed by a stabilizing buoyancy force. An approximate instability condition is readily derived by observing that the formation of a wave (such as that depicted in Figure 6.5) will lead to a reduced pressure, $p_A$, in the gas in the orifice formed by that wave. The reduction below the mean gas pressure, $\bar{p}_G$, will be given by Bernoulli's equation as

$$p_A - \bar{p}_G = -\rho_G u_G^2 a / H \tag{6.6}$$

provided $a \ll H$. The restraining pressure is given by the buoyancy effect of the elevated interface, namely, $(\rho_L - \rho_G)ga$. It follows that the flow will become unstable when

$$u_G^2 > gH\Delta\rho/\rho_G \tag{6.7}$$

In this case, the liquid velocity has been neglected because it is normally small compared with the gas velocity. Consequently, the instability criterion provides an upper limit on the gas velocity that is, in effect, the velocity difference. Taitel and Dukler (1976) compared this prediction for the boundary of the stratified flow regime in a horizontal pipe with the experimental observations of Mandhane et al. (1974) and found substantial agreement.

As a second example, consider vertical annular flow that becomes unstable when the Bernoulli force overcomes the stabilizing surface tension force. From Equation 6.4, this implies that disturbances with wavelengths greater than a critical value, $\lambda_c$, will be unstable and that

$$\lambda_c = 2\pi \mathcal{S}(\rho_1 + \rho_2)\big/\rho_1\rho_2(\Delta u)^2 \tag{6.8}$$

For a liquid stream and a gas stream (as is normally the case in annular flow) and with $\rho_L \ll \rho_G$, this becomes

$$\lambda_c = 2\pi \mathcal{S}\big/\rho_G(\Delta u)^2 \tag{6.9}$$

Now consider the application of this criterion to a well-developed annular flow at high gas volume fraction in which $\Delta u \approx j_G$. Then for a water–air mixture, Equation 6.9 predicts critical wavelengths of 0.4 cm and 40 cm for $j_G = 10$ m/s and $j_G = 1$ m/s, respectively. In other words, at low values of $j_G$, only larger wavelengths are unstable, and this seems to be in accord with the breakup of the flow into large slugs. Conversely, at higher $j_G$ flow rates, even quite small wavelengths are unstable, and the liquid gets torn apart into the small droplets carried in the core gas flow.

## 6.3 Pressure Drop

### 6.3.1 Introduction

An obvious objective of the analysis of the flow in the primary coolant loop is the prediction and understanding of the pressure drop in the flow through the core and the corresponding pressure increase in the flow through the primary coolant pumps. As long as these remain single-phase flow, the analyses do not differ greatly from the parallel features in any power plant, and it will be assumed herein that the reader has some familiarity with such single-phase flow analyses. However, when boiling occurs either by design or because of some abnormal excursion, the resulting multiphase flow requires more complicated analyses, and those methods are briefly reviewed in the next few sections. It should be noted that the literature contains a plethora of engineering correlations for multiphase flow pipe friction and some data for other components, such as pumps. This section provides an overview and some references to illustrative material but does not pretend to survey these empirical methodologies.

### 6.3.2 Horizontal Disperse Flow

As might be expected, frictional losses in straight uniform pipe flows have been the most widely studied, and so it is appropriate to begin with a discussion of that subject, focusing first on disperse or nearly disperse flows and then on separated flows.

Beginning with disperse horizontal flow, it is noted that there exists a substantial body of data relating to the frictional losses or pressure gradient, $(-dp/ds)$, in a straight pipe of circular cross section (the coordinate $s$ is measured along the axis of the pipe). Clearly $(-dp/ds)$ is a critical factor in the design of many systems (e.g., slurry pipelines). This pressure gradient is usually nondimensionalized using the pipe diameter, $d$, the density of the continuous phase ($\rho_C$), and either the total volumetric flux, $j$, or the volumetric flux of the continuous fluid ($j_C$). Thus commonly used friction coefficients are

$$C_f = \frac{d}{2\rho_C j_C^2}\left(-\frac{dp}{ds}\right) \quad \textbf{or} \quad C_f = \frac{d}{2\rho_C j^2}\left(-\frac{dp}{ds}\right) \tag{6.10}$$

and, in parallel with the traditional Moody diagram for single-phase flow, these friction coefficients are usually presented as functions of a Reynolds number for various mixture ratios as characterized by the volume fraction, $\alpha$, or the volume quality, $\beta$, of the disperse phase. Commonly used Reynolds numbers are based on the pipe diameter, the viscosity of the continuous phase ($\nu_C$), and either the total volumetric flux, $j$, or the volumetric flux of the continuous phase, $j_C$. For boiling flows or for gas–liquid flows, the reader is referred to the reviews of Hsu and Graham (1976) and Collier and Thome (1994). For a review of slurry pipeline data, the reader is referred to Shook and Roco (1991) and Lazarus and Neilson (1978). For the solids–gas flows associated with the pneumatic conveying of solids, Soo (1983) provides a good summary.

### 6.3.3 Homogeneous Flow Friction

When the multiphase flow or slurry is thoroughly mixed, the pressure drop can be approximated by the friction coefficient for a single-phase flow with the mixture density, $\rho$ (Equation 6.1), and the same total volumetric flux, $j$, as the multiphase flow. Then the ratio of the multiphase flow friction coefficient (based on $j$), $C_f(\alpha)$, at a particular void fraction, $\alpha$, to the friction coefficient for the continuous phase flowing alone, $C_f(0)$, will given by

$$\frac{C_f(\alpha)}{C_f(0)} = \frac{(1+\alpha\rho_D/\rho_C)}{(1-\alpha)^2} \tag{6.11}$$

where it is assumed that $\beta \approx \alpha$. An example of the comparison of this expression with measured friction coefficient ratios in horizontal disperse flows shows good agreement up to large volume fractions (Brennen 2005).

Thus a flow regime that is homogeneous or thoroughly mixed can usually be modeled as a single-phase flow with an effective density, volume flow rate, and viscosity. In these circumstances, the orientation of the pipe appears to make little difference. Often these correlations also require an effective mixture viscosity. In the preceding example, an effective kinematic viscosity of the multiphase flow could have been incorporated into the expression 6.11; however, this often has little effect, especially under the turbulent conditions.

Wallis (1969) includes a discussion of homogeneous flow friction correlations for both laminar and turbulent flow. Turbulence in multiphase flows introduces another set of complicated issues. Nevertheless, the earlier mentioned single-phase approach to the pipe friction seems to produce moderately accurate results in homogeneous flows, as is illustrated by the data of Figure 6.6. The presence of drops, bubbles, or

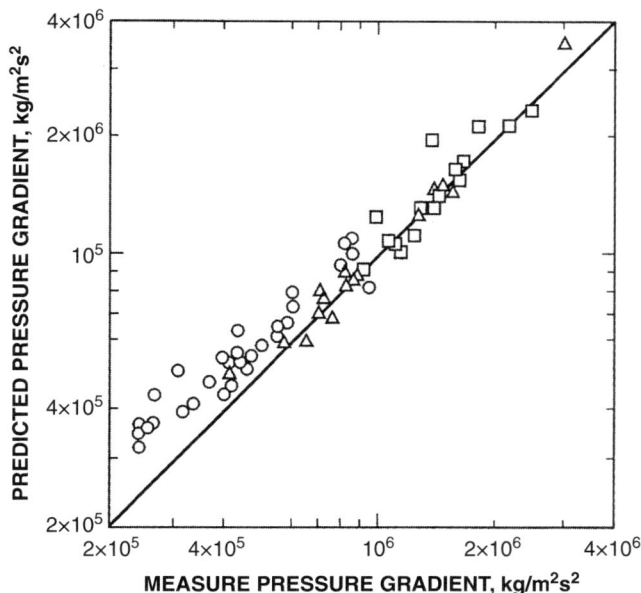

Figure 6.6. Comparison of the homogeneous prediction with measured friction coefficients in a 0.3-cm-diameter tube for steam–water flows with mass qualities, $\mathcal{X}$, ranging up to 0.5. Data from Owens (1961).

particles can act like surface roughness, enhancing turbulence in many applications. Consequently, turbulent friction factors for homogeneous flow tend to be similar to the values obtained for single-phase flow in rough pipes, values around 0.005 being commonly experienced (Wallis 1969).

Vertically oriented pipe flow can experience partially separated flows in which large relative velocities develop due to buoyancy and the difference in the densities of the two phases or components. These large relative velocities complicate the problem of evaluating the pressure gradient and can lead to friction coefficients much larger than suggested by a homogeneous flow friction factor.

### 6.3.4 Frictional Loss in Separated Flow

Having discussed homogeneous and disperse flows, attention is now turned to the friction in separated flows and, in particular, the commonly used Martinelli correlations. The Lockhart–Martinelli (Lockhart and Martinelli 1949) and Martinelli–Nelson (Martinelli and Nelson 1948) correlations are widely documented in multiphase flow texts (see, e.g., Wallis 1969; Brennen 2005). These attempt to predict the frictional pressure gradient in two-component or two-phase pipe flows. It is assumed that these flows consist of two separate co-current streams that, for convenience, are referred to as the liquid and the gas, though they could be any two immiscible fluids. The correlations use the results for the frictional pressure gradient in single-phase pipe flows of each of the two fluid streams. In two-phase flow, the volume fraction is often changing as the mixture progresses along the pipe, and such phase change necessarily implies acceleration or deceleration of the fluids. Associated with this acceleration is an additional acceleration component of the pressure gradient that is addressed with the Martinelli–Nelson correlation. Obviously, it is convenient to begin with the simpler, two-component case (the Lockhart–Martinelli correlation); this also neglects the effects of changes in the fluid densities with distance, $s$, along the pipe axis so that the fluid velocities also remain invariant with $s$. Moreover, in all cases, it is assumed that the hydrostatic pressure gradient has been accounted for so that the only remaining contribution to the pressure gradient, $-dp/ds$, is that due to the wall shear stress, $\tau_w$. A simple balance of forces requires that

$$-\frac{dp}{ds} = \frac{P}{A}\tau_w \tag{6.12}$$

where $P$ and $A$ are the perimeter and cross-sectional area of the stream or pipe. For a circular stream or pipe, $P/A = 4/d$, where $d$ is the stream–pipe diameter. For noncircular cross sections, it is convenient to define a *hydraulic diameter*, $4A/P$. Then, defining the dimensionless friction coefficient, $C_f$, as

$$C_f = \tau_w / \frac{1}{2}\rho j^2 \tag{6.13}$$

the more general form of Equation 6.10 becomes

$$-\frac{dp}{ds} = C_f \rho j^2 \frac{P}{2A} \tag{6.14}$$

In single-phase flow, the coefficient, $C_f$, is a function of the Reynolds number, $\rho d j/\mu$, of the form

$$C_f = K \left\{ \frac{\rho d j}{\mu} \right\}^{-m} \tag{6.15}$$

where $K$ is a constant that depends on the roughness of the pipe surface and will be different for laminar and turbulent flow. The index, $m$, is also different, being 1 in the case of laminar flow and 1/4 in the case of turbulent flow.

These relations from single-phase flow are applied to the two co-current streams in the following way. First, hydraulic diameters, $d_L$ and $d_G$, will be defined for each of the two streams, and the corresponding area ratios, $\kappa_L$ and $\kappa_G$, are then given by

$$\kappa_L = 4A_L/\pi d_L^2 \qquad \kappa_G = 4A_G/\pi d_G^2 \tag{6.16}$$

where $A_L = A(1 - \alpha)$ and $A_G = A\alpha$ are the actual cross-sectional areas of the two streams. The quantities $\kappa_L$ and $\kappa_G$ are shape parameters that depend on the geometry of the flow pattern. In the absence of any specific information on this geometry, one might choose the values pertinent to streams of circular cross section, namely, $\kappa_L = \kappa_G = 1$, and the commonly used form of the Lockhart–Martinelli correlation employs these values. (Note that Brennen 2005 also presents results for an alternative choice.)

The basic geometric relations yield

$$\alpha = 1 - \kappa_L d_L^2/d^2 = \kappa_G d_G^2/d^2 \tag{6.17}$$

Then, the pressure gradient in each stream is assumed given by the following coefficients taken from single-phase pipe flow:

$$C_{fL} = \mathcal{K}_L \left\{ \frac{\rho_L d_L u_L}{\mu_L} \right\}^{-m_L} \qquad C_{fG} = \mathcal{K}_G \left\{ \frac{\rho_G d_G u_G}{\mu_G} \right\}^{-m_G} \tag{6.18}$$

and, because the pressure gradients must be the same in the two streams, this imposes the following relation between the flows:

$$-\frac{dp}{ds} = \frac{2\rho_L u_L^2 \mathcal{K}_L}{d_L} \left\{ \frac{\rho_L d_L u_L}{\mu_L} \right\}^{-m_L} = \frac{2\rho_G u_G^2 \mathcal{K}_G}{d_G} \left\{ \frac{\rho_G d_G u_G}{\mu_G} \right\}^{-m_G} \tag{6.19}$$

In the preceding, $m_L$ and $m_G$ are 1 or 1/4, depending on whether the stream is laminar or turbulent.

Equations 6.17 and 6.19 are the basic relations used to construct the Lockhart–Martinelli correlation. The solutions to these equations are normally and most conveniently presented in nondimensional form by defining the following dimensionless pressure gradient parameters:

$$\phi_L^2 = \frac{\left( \frac{dp}{ds} \right)_{\text{actual}}}{\left( \frac{dp}{ds} \right)_L} \qquad \phi_G^2 = \frac{\left( \frac{dp}{ds} \right)_{\text{actual}}}{\left( \frac{dp}{ds} \right)_G} \tag{6.20}$$

where $(dp/ds)_L$ and $(dp/ds)_G$ are, respectively, the hypothetical pressure gradients that would occur in the same pipe if only the liquid flow were present and if only the gas flow were present. The ratio of these two hypothetical gradients, $Ma$, given by

$$Ma^2 = \frac{\phi_G^2}{\phi_L^2} = \frac{\left( \frac{dp}{ds} \right)_L}{\left( \frac{dp}{ds} \right)_G} = \frac{\rho_G j_G^2}{\rho_L j_L^2} \frac{\mathcal{K}_G}{\mathcal{K}_L} \frac{\left\{ \frac{\rho_G j_G d}{A \mu_G} \right\}^{-m_G}}{\left\{ \frac{\rho_L j_L d}{A \mu_L} \right\}^{-m_L}} \tag{6.21}$$

has come to be called the Martinelli parameter and allows presentation of the solutions to Equations 6.17 and 6.19 in a convenient parametric form. Using the definitions of Equations 6.20, the nondimensional forms of Equations 6.17 become

$$\alpha = 1 - \kappa_L^{(3-m_L)/(m_L-5)} \phi_L^{4/(m_L-5)} = \kappa_G^{(3-m_G)/(m_G-5)} \phi_G^{4/(m_G-5)} \tag{6.22}$$

and the solution of these equations produces the Lockhart–Martinelli prediction of the nondimensional pressure gradient.

To summarize, for given values of (1) the fluid properties, $\rho_L$, $\rho_G$, $\mu_L$, and $\mu_G$; (2) the nature of the flow, laminar or turbulent, in the two streams and the phase correlation constants, $m_L$, $m_G$, $\mathcal{K}_L$, and $\mathcal{K}_G$; (3) the parameters defined by the flow pattern geometry, $\kappa_L$ and $\kappa_G$; and (4) a given value of $\alpha$ Equations 6.22 can be solved to find the nondimensional solution to the flow, namely, the values of $\phi_L^2$ and $\phi_G^2$. The value of $Ma^2$ also follows, and the rightmost expression in Equation 6.21 then yields a relation between the liquid mass flux, $\rho_L j_L$, and the gas mass flux, $\rho_G j_G$. Thus, if one is also given just **one** mass flux (often this will be the total mass flux, $\dot{m} = \rho_L j_L + \rho_G j_G$), the solution will yield the individual mass fluxes, the mass quality, and other flow properties. Alternatively one could begin the calculation with the

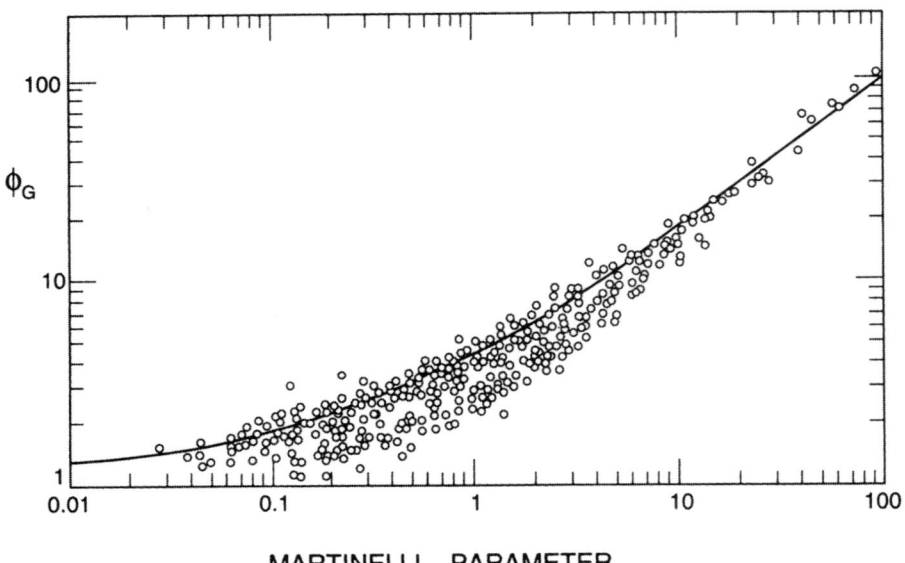

MARTINELLI   PARAMETER

Figure 6.7. Comparison of the Lockhart–Martinelli correlation (the $TT$ case) for $\phi_G$ (solid line) with experimental data. Adapted from Turner and Wallis (1965).

mass quality rather than the void fraction and find the void fraction as one of the results. Finally, the pressure gradient, $dp/ds$, follows from the values of $\phi_L^2$ and $\phi_G^2$.

Charts for the results are presented by Wallis (1969), Brennen (2005), and others. Charts like these are commonly used in the manner described earlier to obtain solutions for two-component gas–liquid flows in pipes. A typical comparison of the Lockhart–Martinelli prediction with the experimental data is presented in Figure 6.7. Note that the scatter in the data is significant (about a factor of 3 in $\phi_G$) and that the Lockhart–Martinelli prediction often yields an overestimate of the friction or pressure gradient. This is the result of the assumption that the entire perimeter of both phases experiences static wall friction. This is not the case, and part of the perimeter of each phase is in contact with the other phase. If the interface is smooth, this could result in a decrease in the friction; conversely, a roughened interface could also result in increased interfacial friction.

It is important to recognize that there are many deficiencies in the Lockhart–Martinelli approach. First, it is assumed that the flow pattern consists of two parallel streams and any departure from this topology could result in substantial errors. Second, there is the previously discussed deficiency regarding the suitability of assuming that the perimeters of both phases experience friction that is effectively equivalent to that of a static solid wall. A third source of error arises because the multiphase flows are often unsteady, and this yields a multitude of quadratic interaction terms that contribute to the mean flow in the same way that Reynolds stress terms contribute to turbulent single-phase flow.

The Lockhart–Martinelli correlation was extended by Martinelli and Nelson (1948) to include the effects of phase change. This extension includes evaluation of the additional pressure gradient due to the acceleration of the flow caused by the phase change. To evaluate this, one must know the variation of the mass quality, $\mathcal{X}$, with distance, $s$, along the pipe. In many boilers, evaporators, or condensers, the rate of heat supply or removal per unit length of the pipe, $\mathcal{Q}_\ell$, is roughly uniform, and the latent heat, $\mathcal{L}$, can be also be considered constant. It follows that for a flow rate of $\dot{m}$ in a pipe of cross-sectional area, $A$, the mass quality varies linearly with distance, $s$, because

$$\frac{d\mathcal{X}}{ds} = \frac{\mathcal{Q}_\ell}{A\dot{m}\mathcal{L}} \tag{6.23}$$

Given the quantities on the right-hand side, this allows evaluation of the mass quality as a function of distance along the conduit and also allows evaluation of the additional acceleration contributions to the pressure gradient. For further details, the reader is referred to Brennen (2005).

## 6.4 Vaporization

### 6.4.1 Classes of Vaporization

There are two classes of rapid vaporization of importance in the context of nuclear reactors, and these are denoted here as homogeneous and heterogeneous vaporization. Homogeneous vaporization occurs when the principal source of the latent heat supply to the interface is the liquid itself. Examples are the formation and growth of a cavitation bubble in a liquid body far from a solid boundary or the vapor explosions described in Section 6.4.3. Conversely, heterogeneous vaporization occurs when the principal source of the latent heat supply to the interface is a different nearby substance or object such as a heated wall. Examples are a pool boiling near a heated surface or many of the fuel–coolant interactions described in Section 7.6.5. Though there is overlap between the two classes, the definitions are convenient in distinguishing the contributing features.

Moreover, each of these two classes can be subdivided into one of two circumstances. The first circumstance is that in which the growth of the vapor volume is only limited by the inertia of the surroundings, liquid or solid. In the second, the vapor volume growth is more severely limited by the rate of supply of latent heat to the interface to produce the vaporization. Both of these rate-limiting growth mechanisms are examined in the sections that follow because the rate of volume growth essentially controls the rate of damage (if any) to the structure in contact with the liquid.

### 6.4.2 Homogeneous Vaporization

Homogeneous vaporization is identified as vaporization in which the principal source of the latent heat supply to the interface is the liquid itself rather than some nearby heat source. A possible model could be a simplified version of the equations governing the dynamics of a simple spherical bubble of radius $R(t)$. The reader who seeks greater detail is referred to the presentation in Brennen (1995) that includes many of the lesser features omitted here. Lord Rayleigh (1917) first derived the equation governing the radius, $R(t)$, of a spherical vapor–gas bubble in a liquid of density, $\rho_L$, when the pressure inside the bubble is $p_B(t)$, the pressure far away in the liquid is $p_\infty(t)$, and the surface tension is $\mathcal{S}$, namely,

$$\frac{p_B(t) - p_\infty(t)}{\rho_L} = R\frac{d^2R}{dt^2} + \frac{3}{2}\left(\frac{dR}{dt}\right)^2 + \frac{2\mathcal{S}}{\rho_L R} \tag{6.24}$$

The pressure in the bubble, $p_B(t)$, may comprise a component due to any noncondensible gas present and the vapor pressure of the surrounding liquid *at the prevailing temperature in the bubble*, $T_B(t)$. The first component, that due to any noncondensible gas, is important but will not be central to the current presentation. Conversely, the vapor pressure and, in particular, the prevailing temperature in the bubble play a key role in the phenomena that are manifest.

In any liquid volume that is mostly at a temperature close to its triple point, the vapor density is so small that only a very small mass of liquid on the surface of the bubble needs to evaporate to supply the increase in bubble volume associated with the bubble growth. Moreover, that small mass of liquid means that only a small supply of heat to the interface is needed to effect the evaporation. And, in turn, that small heat flux only creates a small thermal boundary layer on the bubble surface so that the temperature in the bubble, $T_B(t)$, is only very slightly depressed below the prevailing temperature in the bulk of the liquid, $T_\infty$.

The converse of this is a liquid that is mostly at a higher temperature, so that the density of the vapor is such that a significant mass of liquid must be vaporized at the bubble surface to provide the volume needed for the bubble growth. This implies a substantial heat flux to the interface to provide the latent heat for that evaporation, and that heat flux, in turn, usually causes a significant reduction in the temperature of the bubble contents, $T_B(t)$ (see later for an exception to this consequence). It follows that the vapor pressure in the bubble decreases so that the pressure difference driving the bubble growth, namely, $p_B(t) - p_\infty$, decreases, and therefore, according to Equation 6.24, the rate of bubble growth decreases. This effect of the liquid temperature in depressing the rate of bubble growth is called the *thermal effect on bubble growth*, and it can cause quite a dramatic difference in the resulting bubble dynamics. Perhaps this is most dramatically recognized in the bubble growth in water at normal temperatures. Bubble growth at room temperatures (that are close to the triple point of water) are most frequently observed as cavitation (see Brennen 1995), a phenomenon in which the growth (and the subsequent collapse) of bubbles is extremely explosive and violent. Conversely, bubble growth in a pot of boiling water on the stove at $100°$ C is substantially inhibited by thermal effects and is therefore much less explosive, much less violent.

These effects can be quantified using the following analyses. First, in the case of no thermal effect, the temperature of the bubble contents will be close to the liquid temperature, and therefore the bubble pressure will be roughly constant (neglecting the effect of any noncondensible gas). Then, if the pressures are assumed constant and the surface tension is neglected, Equation 6.24 can be integrated to yield

$$\frac{dR}{dt} = \left[\frac{2(p_B - p_\infty)}{3\rho_L}\right]^{\frac{1}{2}} \qquad R = \left[\frac{2(p_B - p_\infty)}{3\rho_L}\right]^{\frac{1}{2}} t \qquad (6.25)$$

where the integration constant is absorbed into the origin of $t$. This result implies explosive bubble growth, with a volume increasing like $t^3$; it is the kind of bubble growth characteristic of cavitation (Brennen 1995).

For contrast, consider the thermally inhibited growth characteristic of boiling in which the growth is controlled by the rate at which heat can diffuse through an interfacial thermal boundary layer to provide the latent heat of vaporization. The rate of volume growth of the bubble, $4\pi R^2 dR/dt$, requires a mass rate of evaporation equal to $4\pi R^2 (dR/dt)/\rho_V$, where $\rho_V$ is the vapor density in the bubble. To evaporate this mass requires a rate of heat supply to the interface equal to

$$4\pi R^2 (dR/dt)/(\mathcal{L}\rho_V) \qquad (6.26)$$

where $\mathcal{L}$ is the latent heat of evaporation. This heat must diffuse through the thermal boundary layer that builds up in the liquid on the bubble surface and causes the bubble temperature, $T_B$, to fall below the liquid temperature outside of the boundary layer, $T_\infty$. It is this temperature difference, $(T_\infty - T_B)$ that drives heat to the bubble surface at a rate given approximately by

$$4\pi R^2 k_L (T_\infty - T_B)/\delta \qquad (6.27)$$

where $k_L$ is the thermal conductivity of the liquid and $\delta$ is the thickness of the thermal boundary layer. For growth that begins at time $t = 0$, this thickness is given approximately by

$$\delta \approx (\alpha_L t)^{\frac{1}{2}} \qquad (6.28)$$

where $\alpha_L$ is the thermal diffusivity of the liquid and the thinning of the boundary layer as the bubble grows has been neglected. Furthermore, because $T_B$ is the temperature of the interface, it should be roughly equal to the vapor temperature at the bubble pressure, $p_B$, and using the Clausius-Clapeyron relation (Brennen 2005),

$$T_\infty - T_B = \frac{(p_B - p_V)T_B}{\rho_V \mathcal{L}} \qquad (6.29)$$

Equating the expressions 6.26 and 6.27 and using the expressions 6.28 and 6.29, the following bubble growth rate is obtained:

$$\frac{dR}{dt} = \frac{k_L T_B (p_B - p_V)}{(\alpha t)^{\frac{1}{2}}} \quad \text{and so} \quad R \propto t^{\frac{1}{2}} \qquad (6.30)$$

This rate of growth is much slower than given by the expression 6.25 and is characteristic of boiling in water at normal pressures.

To summarize, the preceding analyses (given in much more detail in Brennen 1995) lead, naturally, to two technologically important multiphase phenomena, namely, cavitation and boiling. The essential difference is that bubble growth (and

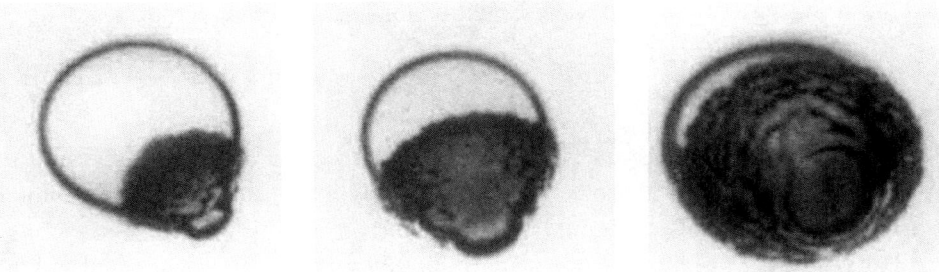

Figure 6.8. Typical photographs of a rapidly growing bubble in a droplet of superheated ether suspended in glycerine. The bubble is the dark, rough mass; the droplet is clear and transparent. The photographs, which are of different events, were taken 31, 44, and 58 $\mu$s after nucleation, and the droplets are approximately 2 mm in diameter. Reproduced from Frost and Sturtevant (1986).

collapse) in boiling is inhibited by limitations on the heat transfer at the interface whereas bubble growth (and collapse) in cavitation is not limited by heat transfer but only by the inertia of the surrounding liquid. Cavitation is therefore an explosive (and implosive) process that is far more violent and damaging than the corresponding bubble dynamics of boiling.

### 6.4.3 Effect of Interfacial Roughness

One of the features that can alter the thermal inhibition of bubble growth occurs when the bubble surface becomes sufficiently roughened to effectively eliminate the thermal boundary layer. This may occur because of an interfacial instability or because of some external interference with the interface. Shepherd and Sturtevant (1982) and Frost and Sturtevant (1986) examined rapidly growing bubbles near the limit of superheat and found growth rates substantially larger than expected when the bubble was in the thermally inhibited range of parameters. Photographs of those bubbles (see Figure 6.8) show that the interface is rough and irregular in places. The enhancement of the heat transfer caused by this roughening is probably responsible for the larger than expected growth rates. Shepherd and Sturtevant (1982) attribute the roughness to the development of a baroclinic interfacial instability. In other circumstances, Rayleigh–Taylor instability of the interface could give rise to a similar effect (Reynolds and Berthoud 1981). A flow with a high turbulence level could have the same consequence, and it seems clear that this suppression of the thermal inhibition plays a key role in the phenomenon of vapor explosions (Section 7.6.4).

## 6.5 Heterogeneous Vaporization

### 6.5.1 Pool Boiling

Attention is now shifted to the heat transfer phenomena associated with heterogeneous vaporization and begins with the most common version of this, namely, pool boiling, in which the vapor bubbles form and grow as a result of the conduction of heat through a bounding solid surface (in a nuclear reactor the surface of the fuel rods). The most obvious application of this information is the boiling that occurs in a BWR. The heat flux per unit area through the solid surface is denoted by $\dot{q}$; the wall temperature is denoted by $T_w$ and the bulk liquid temperature by $T_b$ (or $T_L$). The temperature difference $\Delta T = T_w - T_b$ is a ubiquitous feature of all these problems. Moreover, in almost all cases, the pressure differences within the flow are sufficiently small that the saturated liquid–vapor temperature, $T_e$, can be assumed uniform. Then, to a first approximation, boiling at the wall occurs when $T_w > T_e$ and $T_b \leq T_e$. The label *subcooled boiling* refers to the circumstances when $T_b < T_e$, and the liquid must be heated to $T_e$ before bubbles occur. Conversely, vapor condensation at the wall occurs when $T_w < T_e$ and $T_b \geq T_e$. The label *superheated condensation* refers to the circumstances in which $T_b > T_e$, and the vapor must be cooled to $T_e$ before liquid appears at the wall.

The solid surface may be a plane vertical or horizontal containing surface, or it may be the interior or exterior of a conduit. Another factor influencing the phenomena is whether there is a substantial fluid flow (convection) parallel to the solid surface. For some of the differences between these various geometries and imposed flow conditions, the reader is referred to texts such as Collier and Thome (1994), Hsu and Graham (1976), or Whalley (1987). The next section includes a review of the phenomena associated with a plane horizontal boundary with no convection. Later sections deal with vertical surfaces.

### 6.5.2 Pool Boiling on a Horizontal Surface

Perhaps the most common configuration, known as *pool boiling*, occurs when a pool of liquid is heated from below through a horizontal surface. For present purposes, it will be assumed that the heat flux, $\dot{q}$, is uniform. A uniform bulk temperature far from the wall is maintained because the mixing motions generated by natural convection (and, in boiling, by the motions of the bubbles) mean that most of the liquid is at a fairly uniform time-averaged temperature. In other words, the time-averaged temperature difference, $\Delta T$, occurs within a thin layer next to the wall.

In pool boiling, the relation between the heat flux, $\dot{q}$, and $\Delta T$ is as sketched in Figure 6.9, and events develop with increasing $\Delta T$, as follows. When the pool as a whole has been heated to a temperature close to $T_e$, the onset of nucleate boiling occurs. Bubbles form at nucleation sites on the wall and grow to a size at which the buoyancy force overcomes the surface tension forces acting at the line of attachment of the bubble to the wall. The bubbles then break away and rise through the liquid.

In a steady state process, the vertically upward heat flux, $\dot{q}$, should be the same at all elevations above the wall. Close to the wall, the situation is complex, for several mechanisms increase the heat flux above that for pure conduction through the liquid. First, the upward flux of vapor away from the wall must be balanced by an equal downward mass flux of liquid, and this brings cooler liquid into closer proximity to the wall. Second, the formation and movement of the bubbles enhance mixing in the

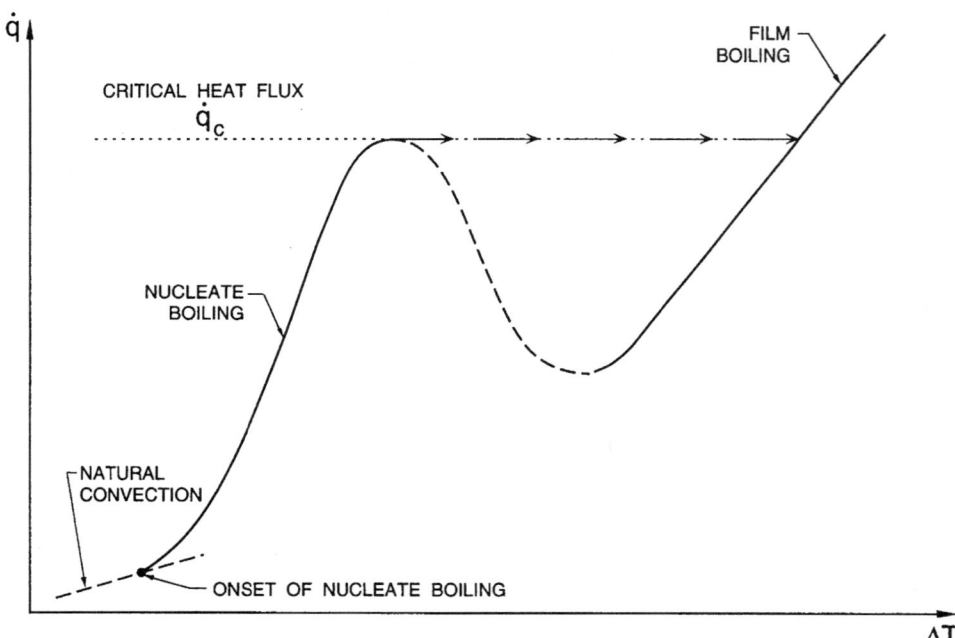

Figure 6.9. Pool boiling characteristics.

liquid near the wall and thus increase heat transfer from the wall to the liquid. Third, the flux of heat to provide the latent heat of vaporization that supplies vapor to the bubbles increases the total heat flux. While a bubble is still attached to the wall, vapor may be formed at the surface of the bubble closest to the wall and then condense on the surface furthest from the wall, thus creating a heat pipe effect. This last mode of heat transfer is sketched in Figure 6.10 and requires the presence of a thin layer of liquid under the bubble known as the *microlayer*.

At distances farther from the wall (Figure 6.11), the dominant component of $\dot{q}$ is simply the enthalpy flux difference between the upward flux of vapor and the downward flux of liquid. Assuming this enthalpy difference is given approximately by the latent heat, $\mathcal{L}$, it follows that the upward volume flux of vapor, $j_V$, is given by $\dot{q}/\rho_V\mathcal{L}$, where $\rho_V$ is the saturated vapor density at the prevailing pressure. Because mass must be conserved, the downward mass flux of liquid must be equal to the upward mass

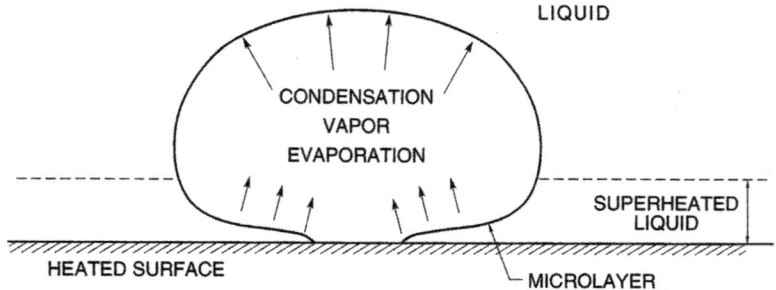

Figure 6.10. Sketch of nucleate boiling bubble with microlayer.

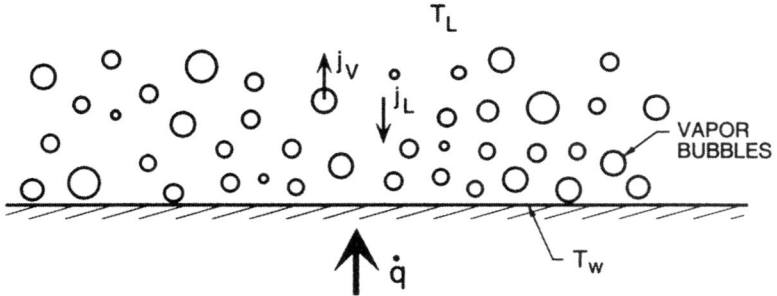

Figure 6.11. Nucleate boiling.

flux of vapor, and it follows that the downward liquid volume flux should be $\dot{q}/\rho_L \mathcal{L}$, where $\rho_L$ is the saturated liquid density at the prevailing pressure.

To complete the analysis, estimates are needed for the number of nucleation sites per unit area of the wall ($N^* \ m^{-2}$), the frequency ($f$) with which bubbles leave each site, and the equivalent volumetric radius ($R$) upon departure. Given the upward velocity of the bubbles ($u_V$), this allows evaluation of the volume fraction and volume flux of vapor bubbles from

$$\alpha = \frac{4\pi R^3 N^* f}{3u_V} \qquad j_V = \frac{4}{3}\pi R^3 N^* f \qquad (6.31)$$

and it then follows that

$$\dot{q} = \frac{4}{3}\pi R^3 N^* f \rho_V \mathcal{L} \qquad (6.32)$$

As $\Delta T$ is increased, both the site density $N^*$ and the bubble frequency $f$ increase until, at a certain critical heat flux, $\dot{q}_c$, a complete film of vapor blankets the wall. This is termed *boiling crisis*, and the heat flux at which it occurs is termed the *critical heat flux (CHF)*. Normally one is concerned with systems in which the heat flux rather than the wall temperature is controlled, and, because the vapor film provides a substantial barrier to heat transfer, such systems experience a large increase in the wall temperature when the boiling crisis occurs. This development is sketched in Figure 6.9. The large increase in wall temperature can be very hazardous, and it is therefore important to be able to predict the boiling crisis and the heat flux at which this occurs. There are a number of detailed analyses of the boiling crisis, and for such detail, the reader is referred to Zuber et al. (1959, 1961), Rohsenow and Hartnett (1973), Hsu and Graham (1976), Whalley (1987), or Collier and Thome (1994). This important fundamental process is discussed in Section 6.5.4.

### 6.5.3 Nucleate Boiling

As Equation 6.32 illustrates, quantitative understanding and prediction of nucleate boiling require detailed information on the quantities $N^*$, $f$, $R$, and $u_V$ and thus knowledge not only of the number of nucleation sites per unit area but also of the cyclic sequence of events as each bubble grows and detaches from a particular site.

Though detailed discussion of the nucleation sites is beyond the scope of this book, it is well established that increasing $\Delta T$ activates increasingly smaller (and therefore more numerous) sites (Griffith and Wallis 1960) so that $N^*$ increases rapidly with $\Delta T$. The cycle of events at each nucleation site as bubbles are created, grow, and detach is termed the *ebullition cycle* and consists of

1. a period of bubble growth during which the bubble growth rate is directly related to the rate of heat supply to each site, $\dot{q}/N^*$. In the absence of inertial effects, and assuming that all this heat is used for evaporation (in a more precise analysis, some fraction is used to heat the liquid), the bubble growth rate is then given by

$$\frac{dR}{dt} = CR^{-2}\frac{\dot{q}}{4\pi\rho_V \mathcal{L}N^*} \tag{6.33}$$

   where $C$ is some constant that will be influenced by complicating factors such as the geometry of the bubble attachment to the wall and the magnitude of the temperature gradient in the liquid normal to the wall (see, e.g., Hsu and Graham 1976).
2. the moment of detachment when the upward buoyancy forces exceed the surface tension forces at the bubble–wall contact line. This leads to a bubble size, $R_d$, upon detachment given qualitatively by

$$R_d = C^*\left[\frac{\mathcal{S}}{g(\rho_L - \rho_V)}\right]^{\frac{1}{2}} \tag{6.34}$$

   where the constant $C^*$ will depend on surface properties such as the contact angle but is of the order of 0.005 (Fritz 1935). With the growth rate from the growth phase analysis, this fixes the time for growth.
3. the waiting period during which the local cooling of the wall in the vicinity of the nucleation site is diminished by conduction within the wall surface and after which the growth of another bubble is initiated.

Obviously, the sum of the growth time and the waiting period leads to the bubble frequency, $f$. In addition, the rate of rise of the bubbles, $u_V$, must be estimated using the methods such as those described in Brennen (2005); note that the downward flow of liquid must also be taken into account in evaluating $u_V$.

These are the basic elements involved in characterizing nucleate boiling though there are many details for which the reader is referred to the texts by Rohsenow and Hartnett (1973), Hsu and Graham (1976), Whalley (1987), or Collier and Thome (1994). Note that the concepts involved in the analysis of nucleate boiling on an inclined or vertical surface do not differ greatly. The addition of an imposed flow velocity parallel to the wall will alter some details because, for example, the analysis of the conditions governing bubble detachment must include consideration of the resulting drag on the bubble.

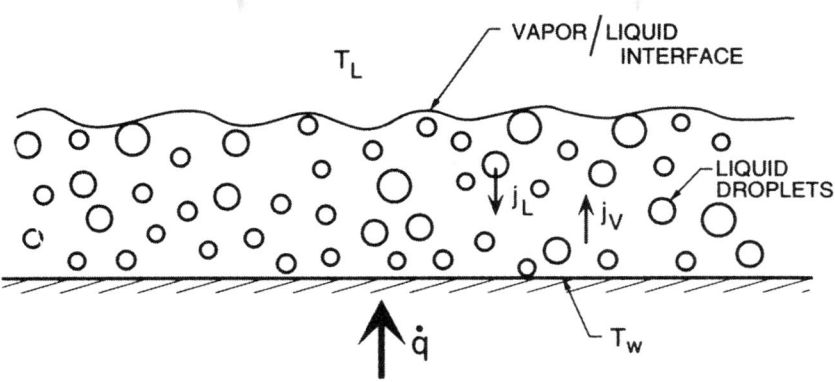

Figure 6.12. Sketch of the conditions close to film boiling.

### 6.5.4 Pool Boiling Crisis

In this section, the approach taken by Zuber et al. (1961) will be followed. They demonstrated that the phenomenon of boiling crisis can be visualized as a flooding phenomenon (see, e.g., Brennen 2005). Consider first the nucleate boiling process depicted in Figure 6.11. As liquid is turned to vapor at or near the solid surface, this results in an upward flux of vapor in the form or bubbles and, necessarily, an equal downward mass flux of liquid. As the heat transfer rate increases these two mass fluxes increase proportionately and the interaction force between the two streams increases. This force inhibits the mass flow rate, and there exists a maximum for which this flow pattern cannot sustain any further increase in heat or mass flux. This is known as the flooding point for this flow pattern, and the maximum or critical heat flux, $\dot{q}_{c1}$, can be estimated (see, e.g., Brennen 2005) to be

$$\dot{q}_{c1} = C_1 \rho_V \mathcal{L} \left\{ \frac{\mathcal{S}g(\rho_L - \rho_V)}{\rho_L^2} \right\}^{\frac{1}{4}}$$
(6.35)

where $\mathcal{L}$ is the latent heat, $\mathcal{S}$ is the surface tension, $\rho_L$ and $\rho_V$ are the liquid and vapor densities, and the typical bubble radius, $R$, is estimated to be given by

$$R = \left\{ \frac{3\mathcal{S}}{2g(\rho_L - \rho_V)} \right\}^{\frac{1}{2}}$$
(6.36)

Now consider the alternative flow pattern sketched in Figure 6.12 in which there is a layer of vapor next to the wall. The flow within that vapor film consists of water droplets falling downward through an upward vapor flow. Analysis of the Rayleigh–Taylor instability of the upper surface of that film leads to the conclusion that the size of the droplets is given by a similar expression as Equation 6.36, except that the factor of proportionality is different. Further analysis of the interaction of downward mass flux of droplets flowing through the upward flux of vapor leads to the conclusion that in this flow pattern, there exists a flooding condition with a maximum possible heat flux and mass flow rate. This maximum heat flux, $\dot{q}_{c2}$, can be estimated

(Brennen 2005) to be

$$\dot{q}_{c2} = C_2 \rho_V \mathcal{L} \left\{ \frac{\mathcal{S}g(\rho_L - \rho_V)}{\rho_V^2} \right\}^{\frac{1}{4}} \tag{6.37}$$

where $C_2$ is some other constant of order unity.

The two model calculations presented earlier (and leading, respectively, to critical heat fluxes given by Equations 6.35 and 6.37) allow the following interpretation of the pool boiling crisis. The first model shows that the bubbly flow associated with nucleate boiling will reach a critical state at a heat flux given by $\dot{q}_{c1}$ at which the flow will tend to form a vapor film. However, this film is unstable, and vapor droplets will continue to be detached and fall through the film to wet and cool the surface. As the heat flux is further increased, a second critical heat flux given by $\dot{q}_{c2} = (\rho_L/\rho_V)^{\frac{1}{2}} \dot{q}_{c1}$ occurs beyond which it is no longer possible for the water droplets to reach the surface. Thus, this second value, $\dot{q}_{c2}$, will more closely predict the true boiling crisis limit. Then, the analysis leads to a dimensionless critical heat flux, $(\dot{q}_c)_{nd}$, from Equation 6.37 given by

$$(\dot{q}_c)_{nd} = \frac{\dot{q}_c}{\rho_V \mathcal{L}} \left\{ \frac{\mathcal{S}g(\rho_L - \rho_V)}{\rho_V^2} \right\}^{-\frac{1}{4}} = C_2 \tag{6.38}$$

Kutateladze (1948) had earlier developed a similar expression using dimensional analysis and experimental data; Zuber et al. (1961) placed it on a firm analytical foundation.

Borishanski (1956), Kutateladze (1952), Zuber et al. (1961), and others have examined the experimental data on critical heat flux to determine the value of $(\dot{q}_c)_{nd}$ (or $C_2$) that best fits the data. Zuber et al. (1961) estimate that value to be in the range 0.12–0.15, although Rohsenow and Hartnett (1973) judge that 0.18 agrees well with most data. Figure 6.13 shows that the values from a wide range of experiments with fluids including water, benzene, ethanol, pentane, heptane, and propane all lie within the range 0.10–0.20. In that figure, $(\dot{q}_c)_{nd}$ (or $C_2$) is presented as a function of the Haberman–Morton number, $Hm = g\mu_L^4 (1 - \rho_V/\rho_L)/\rho_L \mathcal{S}^3$, because the appropriate type and size of bubble that is likely to form in a given liquid will be governed by $Hm$ (see, e.g., Brennen 2005).

Lienhard and Sun (1970) showed that the correlation could be extended from a simple horizontal plate to more complex geometries such as heated horizontal tubes in which the typical dimension (e.g., the tube diameter) is denoted by $d$. Explicitly, Lienhard and Sun recommend

$$(\dot{q}_c)_{nd} = 0.061/C^{**} \quad \text{where} \quad C^{**} = d / \left\{ \frac{\mathcal{S}}{g(\rho_L - \rho_V)} \right\}^{\frac{1}{2}} \tag{6.39}$$

where the constant, 0.061, was determined from experimental data; the result 6.39 should be employed when $C^{**} < 2.3$. For very small values of $C^{**}$ (less than 0.24), there is no nucleate boiling regime, and film boiling occurs as soon as boiling starts.

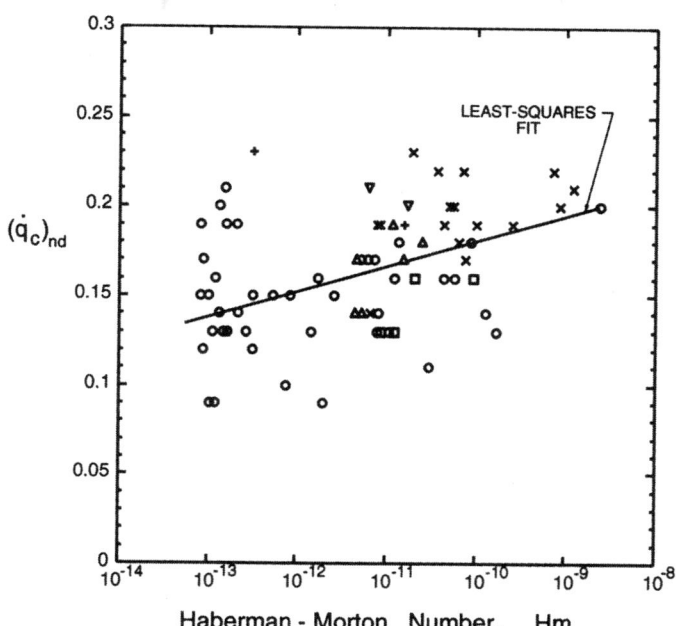

Figure 6.13. Data on the dimensionless critical heat flux, $(\dot{q}_c)_{nd}$ (or $C_2$), plotted against the Haberman–Morton number, $Hm = g\mu_L^4 (1 - \rho_V/\rho_L)/\rho_L S^3$, for water (+), pentane (×), ethanol (□), benzene (△), heptane(▽), and propane (∗) at various pressures and temperatures. Adapted from Borishanski (1956) and Zuber et al. (1961).

For useful reviews of the extensive literature on the critical heat flux in boiling, the reader is referred to Rohsenow and Hartnett (1973), Collier and Thome (1994), Hsu and Graham (1976), and Whalley (1987).

### 6.5.5 Film Boiling

At or near boiling crisis, a film of vapor is formed that coats the surface and substantially impedes heat transfer. This vapor layer presents the primary resistance to heat transfer because the heat must be conducted through the layer. It follows that the thickness of the layer, $\delta$, is given approximately by

$$\delta = \frac{\Delta T k_V}{\dot{q}} \tag{6.40}$$

However, these flows are usually quite unsteady because the vapor–liquid interface is unstable to Rayleigh–Taylor instability (see Section 6.2.5). The result of this unsteadiness of the interface is that vapor bubbles are introduced into the liquid and travel upward while liquid droplets are also formed and fall down through the vapor toward the hot surface. These droplets are evaporated near the surface, producing an upward flow of vapor. The relation 6.40 then needs modification to account for the heat transfer across the thin layer under the droplet.

The droplets do not normally touch the hot surface because the vapor created on the droplet surface nearest the wall creates a lubrication layer that suspends the

Figure 6.14. The evolution of convective boiling around a heated rod. Reproduced from Sherman and Sabersky (1981).

droplet. This is known as the Leidenfrost effect. It is readily observed in the kitchen when a drop of water is placed on a hot plate. Note, however, that the thermal resistance takes a similar form to that in Equation 6.40, although the temperature difference in the vicinity of the droplet now occurs across the much thinner layer under the droplet rather than across the film thickness, $\delta$.

### 6.5.6 Boiling on Vertical Surfaces

Boiling on a heated vertical surface is qualitatively similar to that on a horizontal surface, except for the upward liquid and vapor velocities caused by natural convection. Often this results in a cooler liquid and a lower surface temperature at lower elevations and a progression through various types of boiling as the flow proceeds upward. Figure 6.14 provides an illustrative example. Boiling begins near the bottom of the heated rod, and the bubbles increase in size as they are convected upward. At a well-defined elevation, boiling crisis (Section 6.5.4 and Figure 6.9) occurs and marks the transition to film boiling at a point about five-eighths of the way up the rod in the photograph. At this point, the material of the rod or pipe experiences an abrupt and substantial rise in surface temperature, as described in Section 6.5.2.

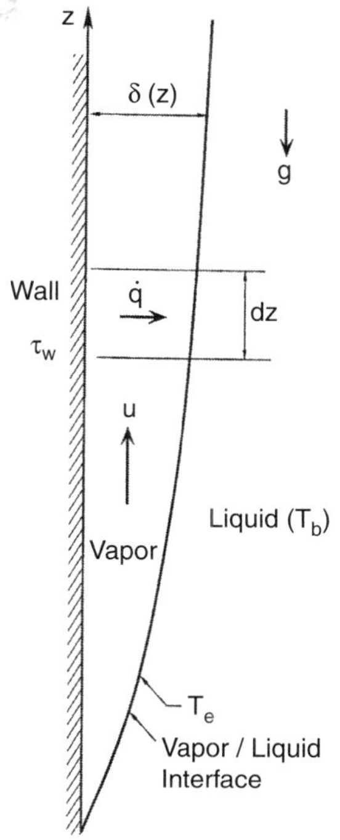

Figure 6.15. Sketch for the film boiling analysis.

The first analysis of film boiling on a vertical surface was due to Bromley (1950) and proceeds as follows. Consider a small element of the vapor layer of length $dz$ and thickness, $\delta(z)$, as shown in Figure 6.15. The temperature difference between the wall and the vapor–liquid interface is $\Delta T$. Therefore the mass rate of conduction of heat from the wall and through the vapor to the vapor–liquid interface per unit surface area of the wall will be given approximately by $k_V \Delta T / \delta$, where $k_V$ is the thermal conductivity of the vapor. In general, some of this heat flux will be used to evaporate liquid at the interface, and some will be used to heat the liquid outside the layer from its bulk temperature, $T_b$, to the saturated vapor–liquid temperature of the interface, $T_e$. If the subcooling is small, the latter heat sink is small compared with the former and, for simplicity in this analysis, it will be assumed that this is the case. Then the mass rate of evaporation at the interface (per unit area of that interface) is $k_V \Delta T / \delta \mathcal{L}$. Denoting the mean velocity of the vapor in the layer by $u(z)$, continuity of vapor mass within the layer requires that

$$\frac{d(\rho_V u \delta)}{dz} = \frac{k_V \Delta T}{\delta \mathcal{L}} \tag{6.41}$$

Assuming that mean values for $\rho_V$, $k_V$, and $\mathcal{L}$ are used, this is a differential relation between $u(z)$ and $\delta(z)$.

A second relation between these two quantities can be obtained by considering the equation of motion for the vapor in the element $dz$. That vapor mass will experience a pressure denoted by $p(z)$ that must be equal to the pressure in the liquid if surface tension is neglected. Moreover, if the liquid motions are neglected so that the pressure variation in the liquid is hydrostatic, it follows that the net force acting on the vapor element as a result of these pressure variations will be $\rho_L g \delta dz$ per unit depth normal to the sketch. Other forces per unit depth acting on the vapor element will be its weight $\rho_V g \delta dz$ and the shear stress at the wall that will be given roughly by $\mu_V u / \delta$. Then, if the vapor momentum fluxes are neglected, the balance of forces on the vapor element yields

$$u = \frac{(\rho_L - \rho_V) g \delta^2}{\mu_V} \tag{6.42}$$

Substituting this expression for $u$ into Equation 6.41 and solving for $\delta(z)$, assuming that the origin of $z$ is chosen to be the origin or virtual origin of the vapor layer where $\delta = 0$, the following expression for $\delta(z)$ is obtained:

$$\delta(z) = \left[ \frac{4 k_V \Delta T \mu_V}{3 \rho_V (\rho_L - \rho_V) g \mathcal{L}} \right]^{\frac{1}{4}} z^{\frac{1}{4}} \tag{6.43}$$

This defines the geometry of the film.

The heat flux per unit surface area of the plate, $\dot{q}(z)$, can then be evaluated and the local heat transfer coefficient, $\dot{q} / \Delta T$, becomes

$$\frac{\dot{q}(z)}{\Delta T} = \left[ \frac{3 \rho_V (\rho_L - \rho_V) g \mathcal{L} k_V^3}{4 \Delta T \mu_V} \right]^{\frac{1}{4}} z^{-\frac{1}{4}} \tag{6.44}$$

Note that this is singular at $z = 0$. It also follows by integration that the overall heat transfer coefficient for a plate extending from $z = 0$ to $z = H$ is

$$\left( \frac{4}{3} \right)^{\frac{3}{4}} \left[ \frac{\rho_V (\rho_L - \rho_V) g \mathcal{L} k_V^3}{\Delta T \mu_V H} \right]^{\frac{1}{4}} \tag{6.45}$$

This characterizes the film boiling heat transfer coefficients in the upper right of Figure 6.9. Though many features of the flow have been neglected, this relation gives good agreement with the experimental observations (Westwater 1958). Other geometrical arrangements such as heated circular pipes on which film boiling is occurring will have a similar functional dependence on the properties of the vapor and liquid (Collier and Thome 1994; Whalley 1987).

## 6.6 Multiphase Flow Instabilities

### 6.6.1 Introduction

Multiphase flows in general are susceptible to a wide range of instabilities over and above those that occur in single-phase flows. A broad review of the state of knowledge of these is beyond the scope of this text. For such a review, the reader is referred to texts such as Brennen (2005). Nevertheless, a brief review of the various types of instability that can occur in multiphase flows is appropriate, and this is followed by some examples that are pertinent to nuclear reactor applications.

It is appropriate to begin by mentioning the basic local instabilities that can occur in these flows. Well known and previously described are some of the local instabilities that can lead to changes in the flow regime, for example, the Kelvin–Helmholtz instability (Section 6.2.5) or boiling crisis (Section 6.5.4).

A second type of instability that can occur can be identified as the system instabilities within an internal flow system that lead to pressure, flow rate, and volume fraction oscillations. These system instabilities can be further subdivided into those that can be analyzed using quasi-static methods (see Brennen 2005), assuming the oscillations progress through a series of quasi-steady states, and, conversely, those that are dynamic. An example of a quasi-static instability is the Ledinegg instability described in Section 6.6.3. An even simpler quasi-static example are the concentration waves that can occur in some circulating systems (Section 6.6.2). However, there are also instabilities that do not have a simple quasi-static explanation and occur in flows that are quasi-statically stable. An example of a fundamentally dynamic instability is the chugging instability described in Section 6.6.4.

### 6.6.2 Concentration Wave Oscillations

Often in multiphase flow processes, one encounters a circumstance in which one part of the flow loop contains a mixture with a concentration that is somewhat different from that in the rest of the system. Such an inhomogeneity may be created during start-up or during an excursion from the normal operating point. It is depicted in Figure 6.16, in which the closed loop has been somewhat arbitrarily divided into a *pipeline* component and a *pump* component. As indicated, a portion of the flow has a mass quality that is larger by $\Delta\mathcal{X}$ than the mass quality in the rest of the system. Such a perturbation could be termed a *concentration wave*, though it is also called a *density wave* or a *continuity wave*; more generally, it is known as a *kinematic wave*. Clearly the perturbation will move around the circuit at a speed that is close to the mean mixture velocity, though small departures can occur in vertical sections in which there is significant relative motion between the phases. The mixing processes that would tend to homogenize the fluid in the circuit are often quite slow so that the perturbation may persist for an extended period.

It is also clear that the pressures and flow rates may vary depending on the location of the perturbation within the system. These fluctuations in the flow variables are termed *concentration wave oscillations*, and they arise from the inhomogeneity of the fluid rather than from any instability in the flow. The characteristic frequency of the oscillations is simply related to the time taken for the flow to complete one circuit of the loop (or some multiple if the number of perturbed fluid pockets is greater than unity). This frequency is usually small, and its calculation often allows identification of the phenomenon.

### 6.6.3 Ledinegg Instability

Sometimes a multiphase flow instability is the result of a nonmonotonic pipeline characteristic. Perhaps the best known example is the Ledinegg instability (Ledinegg 1983) that is depicted in Figure 6.17. This occurs in boiler tubes through which the

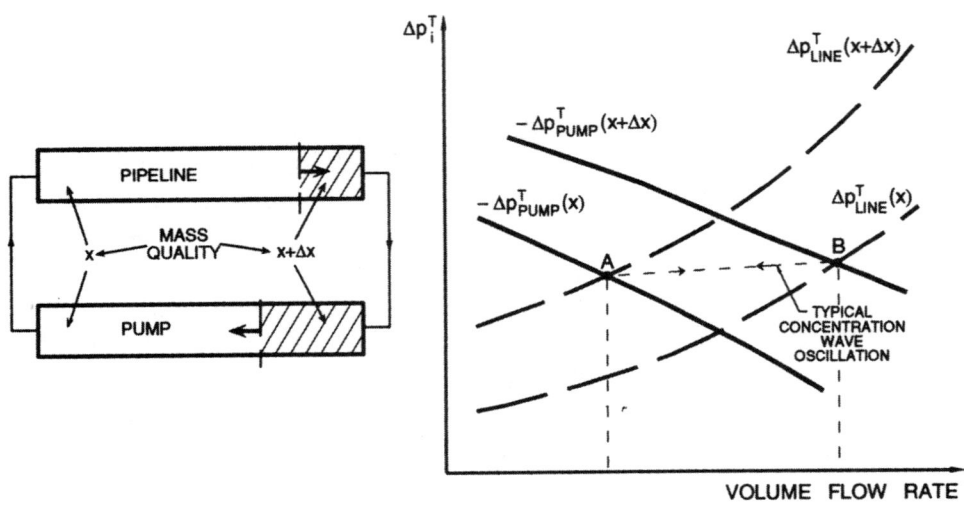

Figure 6.16. Sketch illustrating a concentration wave (density wave) oscillation.

flow is forced either by an imposed pressure difference or by a pump, as sketched in Figure 6.17. If the heat supplied to the boiler tube is roughly independent of the flow rate, then, at high flow rates, the flow will remain mostly liquid because, as discussed in Section 6.2.5, $d\mathcal{X}/ds$ is inversely proportional to the flow rate (see Equation 6.23). Therefore $\mathcal{X}$ remains small. Conversely, at low flow rates, the flow may become mostly vapor because $d\mathcal{X}/ds$ is large. The pipeline characteristic for such a flow (graph of pressure drop versus mass flow rate) is constructed by first considering the two hypothetical characteristics for all-vapor flow and for all-liquid flow. The rough form of these are shown in Figure 6.17; because the frictional losses at high Reynolds numbers are proportional to $\dot{m}^2/\rho$, the all-vapor characteristic lies above the all-liquid line because of the lower density. However, as the flow rate, $\dot{m}$, increases, the actual characteristic must make a transition from the all-vapor line to the all-liquid line and may therefore have the nonmonotonic form sketched in Figure 6.17. Now the system will operate at the point where this characteristic intersects the pump characteristic (or pressure characteristic) driving the flow. This is shown by the solid line(s) in Figure 6.17.

Several examples are shown in Figure 6.17. An operating point such as $A$, where the slope of the pipeline characteristic is greater than the slope of the pump characteristic, will be a stable operating point. This is almost always the case with single-phase flow (see Brennen 2005 for further detail). On the other hand, an operating point such as $B$ is unstable and leads in this example to the Ledinegg instability, in which the operation oscillates back and forth across the unstable region, producing periods of mostly liquid flow interspersed with periods of mostly vapor flow. The instability is most familiar as the phenomenon that occurs in a coffee percolator.

### 6.6.4 Chugging and Condensation Oscillations

An example of a dynamic instability involving a two-phase flow is that which causes the oscillations that occur when steam is forced down a vent into a pool of water.

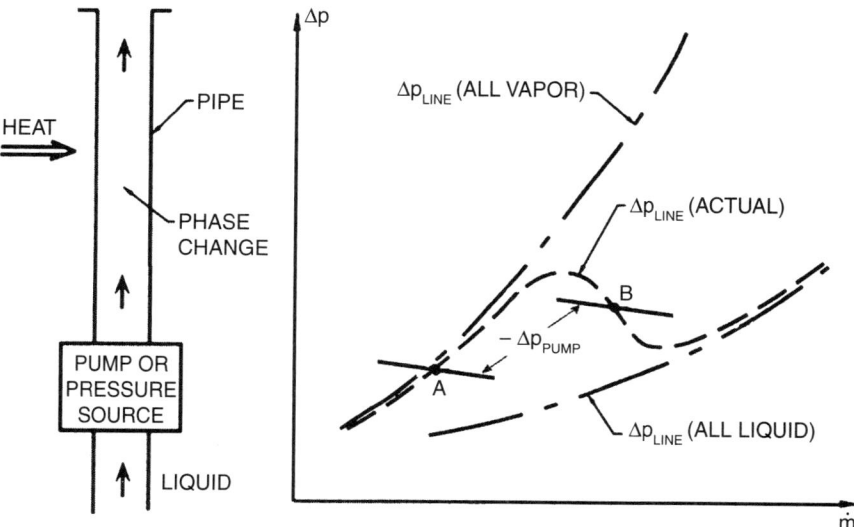

Figure 6.17. Sketch illustrating the Ledinegg instability.

The situation is sketched in Figure 6.18 and is clearly relevant to the pressure suppression systems used in BWRs (see Section 7.4), a context in which the phenomena have been extensively studied (see, e.g., Wade 1974; Koch and Karwat 1976; Class and Kadlec 1976; Andeen and Marks 1978). The phenomena do, however, also occur in other systems in which steam (or other vapor) is injected into a condensing liquid (Kiceniuk 1952). The instabilities that result from the dynamics of a condensation interface can take a number of forms, including those known as *chugging* and *condensation oscillations*.

The basic components of the system are as shown in Figure 6.18 and consist of a vent or pipeline of length $\ell$, the end of which is submerged to a depth, $H$, in a pool of water. The basic instability is illustrated in Figure 6.19. At relatively low steam flow rates, the rate of condensation at the steam–water interface is sufficiently high that the interface remains within the vent. However, at higher flow rates, the pressure in the steam increases and the interface is forced down and out of the end of the vent. When this happens, both the interface area and the turbulent mixing in the vicinity of the interface increase dramatically. This greatly increases the condensation rate that, in turn, causes a marked reduction in the steam pressure. Thus the interface collapses back into the vent, often quite violently. Then the cycle of growth and collapse, of oscillation of the interface from a location inside the vent to one outside the end of the vent, is repeated. The phenomenon is termed *condensation instability* and, depending on the dominant frequency, the violent oscillations are known as *chugging* or *condensation oscillations* (Andeen and Marks 1978).

The frequency of the phenomenon tends to lock in on one of the natural modes of oscillation of the system in the absence of condensation. There are two obvious natural modes and frequencies. The first is the manometer mode of the liquid inside the end of the vent. In the absence of any steam flow, this manometer mode will have a typical small-amplitude frequency, $\omega_m = (g/H)^{\frac{1}{2}}$, where $g$ is the

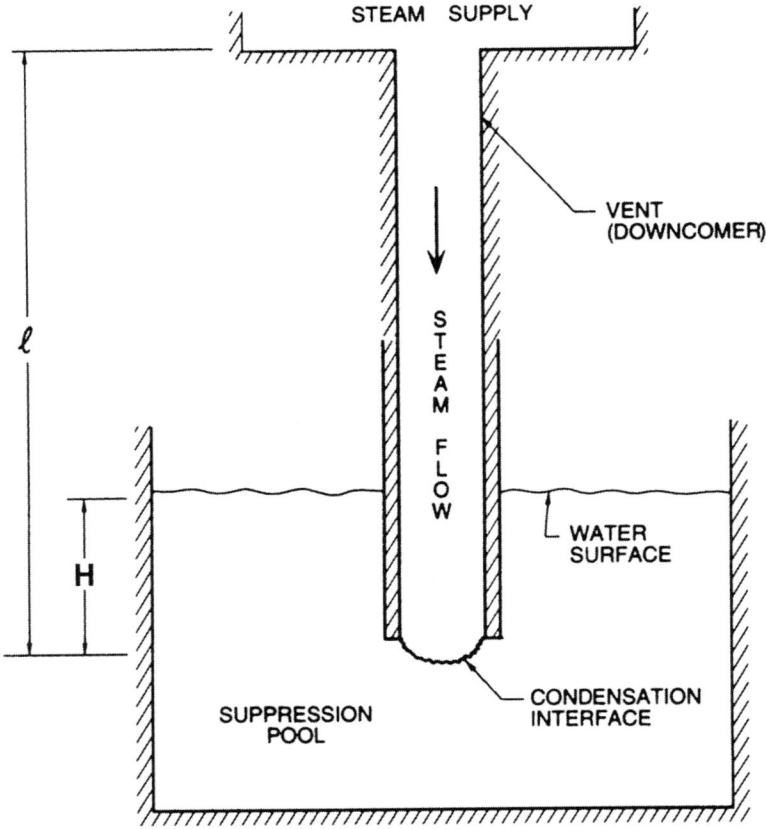

Figure 6.18. Components of a pressure suppression system.

acceleration due to gravity. This is usually a low frequency of the order of 1 Hz or less, and when the condensation instability locks into this low frequency, the phenomenon is known as *chugging*. The pressure oscillations resulting from chugging can be quite violent and can cause structural loads that are of concern to the safety engineer. Another natural mode is the first acoustic mode in the vent whose frequency, $\omega_a$, is approximately given by $\pi c/\ell$, where $c$ is the sound speed in the steam. There are also observations of lock-in to this higher frequency, and these oscillations are known as *condensation oscillations*. They tend to be of smaller amplitude than the chugging oscillations.

Figure 6.20 illustrates the results of a linear stability analysis of the suppression pool system (Brennen 1979). Constructing dynamic transfer functions for each basic component of this system (see Brennen 2005), one can calculate the linearized input impedance of the system viewed from the steam supply end of the vent. In such a linear stability analysis, a positive input resistance implies that the system is absorbing fluctuation energy and is therefore stable; a negative input resistance implies an unstable system. In Figure 6.20, the input resistance is plotted against the perturbation frequency for several steam flow rates. Note that, at low steam flow rates, the system is stable for all frequencies. However, as the steam flow rate is increased,

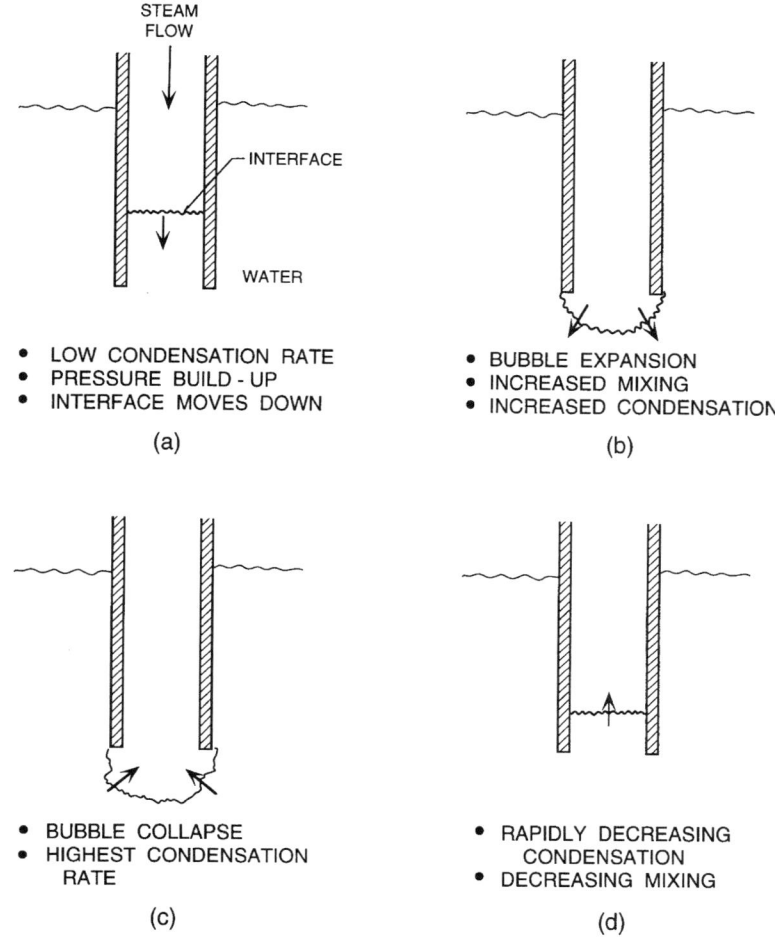

Figure 6.19. Sketches illustrating the stages of a condensation oscillation.

the system first becomes unstable over a narrow range of frequencies close to the manometer frequency, $\omega_m$. Thus chugging is predicted to occur at some critical steam flow rate. At still higher flow rates, the system also becomes unstable over a narrow range of frequencies close to the first vent acoustic frequency, $\omega_a$; thus the possibility of condensation oscillations is also predicted. Note that the quasi-static input resistance at small frequencies remains positive throughout, and therefore the system is quasi-statically stable for all steam flow rates. Thus chugging and condensation oscillations are true, dynamic instabilities.

It is, however, important to observe that a linear stability analysis cannot model the highly nonlinear processes that occur during a *chug* and, therefore, cannot provide information on the subject of most concern to the practical engineer, namely, the magnitudes of the pressure excursions and the structural loads that result from these condensation instabilities. Although models have been developed in an attempt to make these predictions (see, e.g., Sargis et al. 1979), they are usually very specific to the particular problem under investigation. Often, they must also resort to

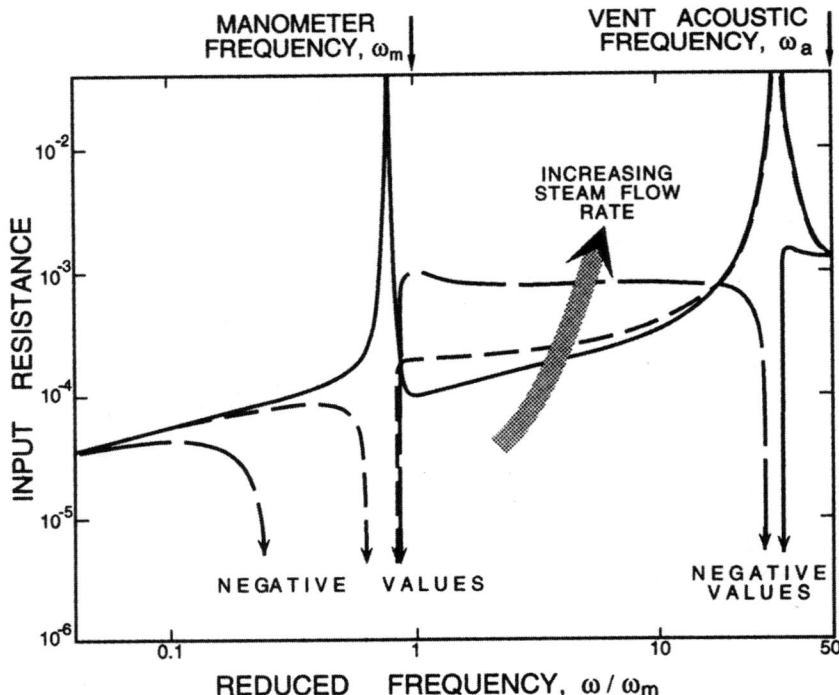

Figure 6.20. The real part of the input impedance (the input resistance) of the suppression pool as a function of the perturbation frequency for several steam flow rates. Adapted from Brennen (1979).

empirical information on unknown factors such as the transient mixing and condensation rates.

Finally, note that these instabilities have been observed in other contexts. For example, when steam was injected into the wake of a streamlined underwater body to explore underwater jet propulsion, the flow became very unstable and oscillated wildly (Kiceniuk 1952).

## 6.7 Nuclear Reactor Context

The next chapter includes descriptions of how multiphase flow is pertinent to the understanding and analysis of nuclear power generation. Multiphase flows arise not only during nominal reactor operation but, even more importantly, in the initiation and development of nuclear reactor accidents.

### REFERENCES

Andeen, G. B., and Marks, J. S. (1978). Analysis and testing of steam chugging in pressure systems. *Electric Power Research Institute* Rep. NP-908.
Baker, O. (1953). Design of pipelines for the simultaneous flow of oil and gas. Society of Petroleum Engineers. doi:10.2118/323-G.
Borishanski, V. M. (1956). An equation generalizing experimental data on the cessation of bubble boiling in a large volume of liquid. *Zhurnal Tekhnicheskoi Fiziki*, **26**, 452–56.

Brennen, C. E. (1979). A linear, dynamic analysis of vent condensation stability. In *Basic mechanisms in two-phase flow and heat transfer* (eds. P. H. Rothe and R. T. Lahey-Jr.). ASME Vol. G00179, 63–71.

Brennen, C. E. (1995). *Cavitation and bubble dynamics.* Oxford University Press.

Brennen, C. E. (2005). *Fundamentals of multiphase flow.* Cambridge University Press.

Bromley, L. A. (1950). Heat transfer in stable film boiling. *Chemical Engineering Progress,* **46**, 221–27.

Butterworth, D., and Hewitt, G. F. (1977). *Two-phase flow and heat transfer.* Oxford University Press.

Class, G., and Kadlec, J. (1976). Survey of the behavior of BWR pressure suppression systems during the condensation phase of LOCA. Paper presented at American Nuclear Society International Conference, Washington, DC.

Collier, J. G., and Thome, J. R. (1994). *Convective boiling and condensation.* Clarendon Press.

Elghobashi, S. E., and Truesdell, G. C. (1993). On the two-way interaction between homogeneous turbulence and dispersed solid particles. I: Turbulence modification. *Physics of Fluids,* **A5**, 1790–1801.

Fritz, W. (1935). The calculation of the maximum volume of steam bladders. *Physikalische Zeitschrift,* **36**, 379–84.

Frost, D., and Sturtevant, B. (1986). Effects of ambient pressure on the instability of a liquid boiling explosively at the superheat limit. *ASME Journal of Heat Transfer,* **108**, 418–24.

Griffith, P., and Wallis, J. D. (1960). The role of surface conditions in nucleate boiling. *Chem. Eng. Prog. Symp.,* Ser. 56, **30**, 49–63.

Hewitt, G. F. (1982). Flow regimes. In *Handbook of multiphase systems* (ed. G. Hetsroni). McGraw-Hill.

Hewitt, G. F., and Hall-Taylor, N. S. (1970). *Annular two-phase flow.* Pergamon Press.

Hewitt, G. F., and Roberts, D. N. (1969). Studies of two-phase flow patterns by simultaneous X-ray and flash photography. *U.K.A.E.A.* Rep. No. AERE-M2159.

Hsu, Y.-Y., and Graham, R. W. (1976). *Transport processes in boiling and two-phase systems.* Hemisphere/McGraw-Hill.

Hubbard, N. G., and Dukler, A. E. (1966). *The characterization of flow regimes in horizontal two-phase flow.* Heat Transfer and Fluid Mechanics Institute, Stanford University.

Jones, O. C., and Zuber, N. (1974). Statistical methods for measurement and analysis of two-phase flow. Paper presented at the International Heat Transfer Conference, Tokyo.

Kiceniuk, T. (1952). A preliminary investigation of the behavior of condensible jets discharging in water. *Calif. Inst. of Tech. Hydro. Lab.* Rep., E-24.6.

Koch, E., and Karwat, H. (1976). Research efforts in the area of BWR pressure suppression containment systems. Paper presented at the 4th Water Reactor Safety Research Meeting, Gaithersburg, MD.

Kutateladze, S. S. (1948). On the transition to film boiling under natural convection. *Kotloturbostroenie,* **3**, 152–158.

Kutateladze, S. S. (1952). Heat transfer in condensation and boiling. *U.S. AEC* Rep. AEC-tr-3770.

Lamb, H. (1932). *Hydrodynamics.* Cambridge University Press.

Lazarus, J. H., and Neilson, I. D. (1978). A generalized correlation for friction head losses of settling mixtures in horizontal smooth pipelines. *Hydrotransport 5,* **1**, B1-1–B1-32.

Ledinegg, M. (1983). Instabilität der Strömung bei Natürlichen und Zwangumlaut. *Warme,* **61**, 891–98.

Lienhard, J. H., and Sun, K. (1970). Peak boiling heat flux on horizontal cylinders. *International Journal of Heat and Mass Transfer,* **13**, 1425–40.

Lockhart, R. W., and Martinelli, R. C. (1949). Proposed correlation of data for isothermal two-phase two-component flow in pipes. *Chemical Engineering Progress,* **45**, 39–48.

Mandhane, J. M., Gregory, G. A., and Aziz, K. A. (1974). A flow pattern map for gas-liquid flow in horizontal pipes. *International Journal of Multiphase Flow,* **1**, 537–53.

Martinelli, R. C., and Nelson, D. B. (1948). Prediction of pressure drop during forced circulation boiling of water. *Transactions of the ASME*, **70**, 695–702.

Owens, W. L. (1961). Two-phase pressure gradient. International Developments in Heat Transfer, **41**(2), **2**, 363–68.

Pan, Y., and Banerjee, S. (1997). Numerical investigation of the effects of large particles on wall-turbulence. *Physics of Fluids*, **9**, 3786–807.

Rayleigh, Lord. (1917). On the pressure developed in a liquid during the collapse of a spherical cavity. *Philosophical Magazine*, **34**, 94–98.

Reynolds, A. B., and Berthoud, G. (1981). Analysis of EXCOBULLE two-phase expansion tests. *Nuclear Engineering and Design*, **67**, 83–100.

Rohsenow, W. M., and Hartnett, J. P. (1973). *Handbook of heat transfer*. Section 13. McGraw-Hill.

Sargis, D. A., Stuhmiller, J. H., and Wang, S. S. (1979). Analysis of steam chugging phenomena, Volumes 1, 2 and 3. *Electric Power Res. Inst.* Rep. NP-962.

Schicht, H. H. (1969). Flow patterns for an adiabatic two-phase flow of water and air within a horizontal tube. *Verfahrenstechnik*, **3**, 153–61.

Shepherd, J. E., and Sturtevant, B. (1982). Rapid evaporation near the superheat limit. *Journal of Fluid Mechanics*, **121**, 379–402.

Sherman, D. C., and Sabersky, R. H. (1981). Natural convection film boiling on a vertical surface. *Letters in Heat and Mass Transfer*, **8**, 145–53.

Shook, C. A., and Roco, M. C. (1991). *Slurry flow: Principles and practice.* Butterworth-Heinemann.

Soo, S. L. (1983). Pneumatic transport. In *Handbook of fluids in motion* (eds. N. P. Cheremisinoff and R. Gupta), Ann Arbor Science.

Squires, K. D., and Eaton, J. K. (1990). Particle response and turbulence modification in isotropic turbulence. *Physics of Fluids*, **A2**, 1191–203.

Taitel, Y., and Dukler, A. E. (1976). A model for predicting flow regime transitions in horizontal and near horizontal gas-liquid flow. *AIChE Journal*, **22**, 47–55.

Turner, J. M., and Wallis, G. B. (1965). The separate-cylinders model of two-phase flow. *AEC* Rep. NYO-3114-6.

Wade, G. E. (1974). Evolution and current status of the BWR containment system. *Nuclear Safety*, **15**, 163–73.

Wallis, G. B. (1969). *One-dimensional two-phase flow.* McGraw-Hill.

Weisman, J. (1983). Two-phase flow patterns. In *Handbook of fluids in motion*, pp. 409–25 (eds. N. P. Cheremisinoff and R. Gupta). Ann Arbor Science.

Weisman, J., and Kang, S. Y. (1981). Flow pattern transitions in vertical and upwardly inclined lines. *International Journal of Multiphase Flow*, **7**, 271–91.

Westwater, J. W. (1958). Boiling of liquids. *Advances in Chemical Engineering*, **2**, 1–56.

Whalley, P. B. (1987). *Boiling, condensation and gas-liquid flow.* Oxford Science.

Yih, C.-S. (1969). *Fluid mechanics.* McGraw-Hill.

Zuber, N. (1959). Hydrodynamic aspects of boiling heat transfer. Ph.D. thesis, UCLA.

Zuber, N., Tribus, M., and Westwater, J. W. (1961). The hydrodynamic crisis in pool boiling of saturated and subcooled liquids. *Proceedings of the 2nd International Heat Transfer Conference*, Section A, Part II, 230–37.

# Reactor Multiphase Flows and Accidents

### 7.1 Multiphase Flows in Nuclear Reactors

This and the following sections will include descriptions of how multiphase flow is pertinent to the understanding and analysis of nuclear power generation and nuclear reactor accidents. The focus is on those multiphase issues that arise in the reactor itself, though, of course, there are many multiphase flow issues associated with the conventional components of the power generation process such as the steam generators and steam turbines.

Multiphase flows that might or do occur in a nuclear reactor are most conveniently subdivided into those that occur during nominal reactor operation and those that might occur and have occurred during a reactor accident. Both sets of issues are complex and multifaceted, and many of the complexities are beyond the scope of this monograph. The reader is referred to texts such as Hsu and Graham (1976), Jones and Bankhoff (1977a, 1977b), Jones (1981), Hewitt and Collier (1987), and Todres and Kazimi (1990) for a broader perspective on these issues.

### 7.1.1 Multiphase Flow in Normal Operation

The most obvious multiphase flow occurring during normal operation is the process of boiling in a BWR core. Sections 6.5.4, 6.5.5, and 6.5.6 described how boiling is initiated within a BWR reactor core (Section 6.5.3), how the flow pattern within the coolant passages would change from bubbly flow to annular flow as the fluid rose (Section 6.2.3), and the circumstances under which the wall film might undergo burnout (Section 6.5.4), leading to the critical heat flux condition (CHF) and a rapid rise in the temperature (Figure 6.9) of the interface between the fuel rod cladding and the coolant. Boiling water reactors are designed to operate at a comfortable margin short of CHF at any location within the reactor. This requires a coupled calculation of the multiphase flow and the neutronics (Section 5.5) as well as a criterion that determines the CHF. For a review of the thermohydraulic data on CHF in nuclear reactors, the reader is referred to Groeneveld and Gardiner (1977).

### 7.1.2 Void Fraction Effect on Reactivity

In most reactors, it is important to recognize that any change in the geometry of the core or change of phase of its components may alter the reactivity of the reactor. Any positive change in the reactivity, $\rho$ (or multiplication factor, $k$), that resulted from an unexpected change in the geometry would clearly be a serious safety issue. Therefore an important objective in the design of a reactor core is to achieve as negative an effect on the reactivity as possible in the event of a change of the geometry of the structure or coolant in the core.

Insofar as the design of the structure of the core (particularly the topological distribution of the fuel, coolant, moderator, etc.) is concerned, the objective is to create an arrangement whose reactivity would decrease in the event of any structural deformation. Some examples are (1) the CANDU reactor design incorporates such an effect (see Section 4.4) and (2) analyses of a hypothetical core disruptive accident in an LMFBR suggest that expansion of the core in a serious accident would also result in a decrease in the reactivity (see Section 7.6.3).

However, perhaps the most important effect in this category occurs in liquid-cooled reactors where any change of phase, any boiling in a PWR or LMFBR, or increased boiling in a BWR can substantially affect the neutronics of the core and the reactivity of the reactor. Thus, one important objective of the multiphase flow analyses of postulated accidents is to assess the *void coefficient*, the change of the reactivity, $\rho$, as a result of a change in the void fraction, $\alpha$, or

$$\text{Void Coefficent} = \frac{d\rho}{d\alpha} \tag{7.1}$$

This may, of course, be a function not only of time and location in the core but of other topological effects. As discussed elsewhere, one of the substantial safety features of water-cooled thermal reactors is that boiling and the loss of coolant that results from overheating cause a strong negative void coefficient because the thermal neutron supply is decreased by the reduction in the moderator (Section 7.4). In contrast, most LMFBR designs have a positive void coefficient (see Section 7.6.3) because the loss of the neutron slowing effect of the sodium coolant results in an increase in the population of fast neutrons. However, modified design of the geometry of the LMFBR core could reverse this dangerous attribute.

### 7.1.3 Multiphase Flow during Overheating

In any light water reactor, it is clear that in the event of any departure from normal operation, whether through unexpected depressurization or through decrease in the coolant flow (e.g., a loss-of-coolant accident, or LOCA), conditions in the reactor core may lead to the critical heat flux (CHF) condition being exceeded with the concominant large increase in the fuel rod temperatures. Such a circumstance could be the precursor for a core meltdown, hence the importance of being able to predict the CHF.

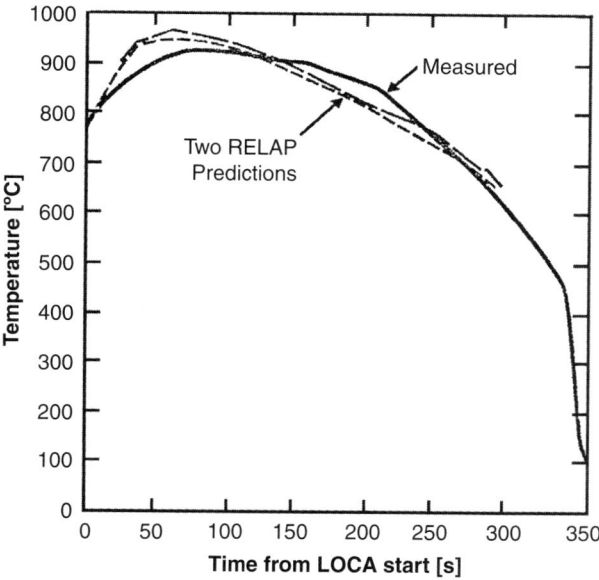

Figure 7.1. An example of a comparison between the measured cladding temperature following a simulated LOCA in a PWR model test facility (FLECHT) and the predictions of the RELAP code with two different choices of coefficients. Adapted from Hsu and Sullivan (1977).

As remarked at the end of Chapter 5 (Section 5.6), the prediction of the flows and temperatures following postulated reactor excursions and accidents is an important input to the evaluation of reactor safety. Much effort has gone into the development and validation of multiphase flow computer codes for this purpose. The objective is to make reliable predictions for the purposes of designing effective safety systems for reactors. An example of the multiphase flow and heat transfer codes developed is the extensively used RELAP code (Aerojet Nuclear Co. 1976; also, e.g., Jackson et al. 1981; Wagner and Ransom (1982). The details of these codes are beyond the scope of this text, and the reader is referred to the references listed later for further information. As with most multiphase numerical methods, validation presents a real challenge for the scaling of many of the phenomena involved contains uncertainties, and the coefficients that govern the flow and heat transfer are hard to predict accurately. Consequently, there is a need for large-scale test facilities and experimental measurements that can be used for validation of these codes. Examples of these facilities and test programs, summarized by Hsu and Sullivan (1977), are the FLECHT program at Westinghouse (see, e.g., Hassan 1986) and the LOFT and other facilities at the Idaho National Engineering Laboratory. Figure 7.1 presents one example of a comparison between a large-scale facility measurement and a computer code. It shows a comparison between a measured cladding temperature in a FLECHT experiment simulating a LOCA and two corresponding predictions using the RELAP code. The discrepancies are typical of the uncertainties in these complex multiphase flow predictions.

## 7.2 Multiphase Flows in Nuclear Accidents

Attention is now switched to nuclear accidents and the safety systems designed and installed to mitigate the consequences of those accidents. Safety concerns are naturally critical to the public acceptance of nuclear power plants, and it is appropriate to review the systems that have been developed and improved to address those concerns. The three major accidents to date, Three Mile Island, Chernobyl, and Fukushima, are briefly described, and the lessons learned from those and other lesser accidents are emphasized because they have led to substantial improvement of the world's nuclear power stations.

## 7.3 Safety Concerns

There are two coupled, major concerns for the designer, manufacturer, and operator of a nuclear power station. The first of these is to avoid any hazard associated with uncontrolled criticality of the reactor, and the second is to eliminate any possible release of radioactive material to the environment surrounding the plant. The designer, manufacturer, and operator seek to minimize the likelihood of any accident, and this requires not only constant vigilance but also continuing improvement in the monitoring instrumentation and in the training of the plant operators.

Over the years, partly because of both the major and lesser accidents that have occurred at nuclear power stations, a great deal of time and effort has gone into examining every conceivable failure (both mechanical and human) that might lead to a departure from controlled operation of a nuclear reactor power station (USAEC 1957, 1973). Fault trees (Bodansky 1996) have been exhaustively explored to try to eliminate any combination of malfunction and/or operator mismanagement that might have serious consequences. Experience, for example during the Three Mile Island accident, has shown that a relatively minor equipment failure combined with human operator error can lead to a serious accident, even to a release of radioactivity.

Moreover, technical and operational analyses must be carried well beyond the initial failure and until a safe and controlled state has again been established. Thus the failure trees must postulate quite unlikely initial failures and then follow the progression of events that necessarily unfold in the seconds, minutes, hours, and weeks that follow. Thus, for example, much attention has been given to the hypothetical LOCA (see Sections 7.6.2 and 7.6.3) that would occur if part of the primary coolant circuit were to fail so that coolant were to escape into the secondary containment and the heat produced in the reactor were no longer being removed by the coolant. The subsequent buildup of heat within the reactor could lead to a meltdown of the core and its containment, a scenario that became popularized by the movie *The China Syndrome* (see, e.g., Lewis 1977; Okrent 1981; Collier and Hewitt 1987). The likelihood that such a meltdown would also lead to a release of radioactive material led to exhaustive study of this particular developing fault path.

These explorations of conceivable fault trees and accidents led to the installation of equipment designed to mitigate the effects of these unlikely events. Indeed, to minimize the potential of human error, it is also desirable that these safety systems be *passive* (not requiring human or mechanical intervention and not requiring power), though this is not always possible. The next section provides information on some of the installed safety systems, with particular focus on those systems designed to mitigate the consequences of a loss of coolant accident.

Another class of concerns is the vulnerability of nuclear reactors to large external events and forces, such as earthquakes, tsunamis, volcanoes, hurricanes, power outages, and terrorist attacks. Many of these involve the choice of the site of a nuclear power plant. Particularly in California, a great deal of attention has been given to the proximity of earthquake faults and the need to ensure that the reactor, its containment structures, and emergency power systems are as impervious as possible to a major earthquake (Bodansky 1996; Okrent 1981). Moreover, these power plants require copious external cooling water and are therefore often sited close to the ocean. The Fukushima accident (see Section 7.5.3) demonstrated that more thought should have been given to protecting the plant and its surrounding auxiliary facilities from the tsunami danger. Another scenario that needed to be examined in the aftermath of the 9/11 disaster in 2001 was the possibility of a direct hit by a fully loaded airliner. Analyses and tests have shown that under no circumstances would there be any penetration of the containment building; the airliner would simply disintegrate.

Of course, public imagination conjures up the possibility of an even more drastic accident, namely, a nuclear explosion. It is, however, contrary to the fundamental laws of physics for any commercial nuclear reactor containing fuel enriched to less than 5 percent to explode like a nuclear bomb. It is also important to emphasize that, apart from the Chernobyl accident, *no one* (neither a member of the public nor a plant worker) has ever died as a result of exposure to a commercial nuclear reactor incident. Moreover, as discussed later, the world has put any future Chernobyl incident beyond the realm of possibility.

## 7.4 Safety Systems

The safety systems installed in modern nuclear reactors for electricity generation have three basic purposes: (1) to control the reactivity of the reactor, to maintain it in a marginally critical state during power production and to shut the reactor down when that is required; (2) to cool the fuel and prevent overheating; and (3) to contain all radioactive substances and radiation even in the event of radical, hypothetical accidents. While a detailed description of each of these strategic objectives is beyond the scope of this text, it is appropriate to comment on each individually.

Though the control of a nuclear power plant is a complex and multifaceted issue (see, e.g., Schultz 1955), the reactivity of a normally operating reactor is primarily controlled by the insertion or withdrawal of the control rods whose effect was demonstrated in Section 3.7.4. One of the most reassuring features of water-cooled and moderated nuclear reactors (in effect, most of the present commercial reactors) is that any overheating of the core that is sufficient to vaporize the cooling water within it will automatically result in a decrease in the reactivity (because thermal neutrons are not

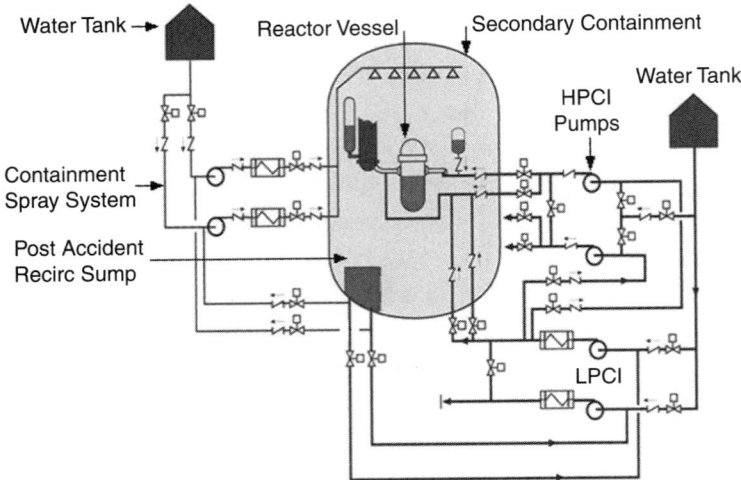

Figure 7.2. Schematic of the ECCS system in a PWR.

being fed back to the fuel) and consequently a shutdown of the nuclear reactor core. Of course, the fuel will still produce decay heat, and therefore special cooling systems are needed to prevent the decay heat from causing an excessive overheating of the core.

Consequently, all modern nuclear reactors are equipped with redundant emergency core cooling systems (ECCS) that force cooling water into the primary containment vessel and the core in the event of an uncontrolled buildup of heat. Some of these systems are passive (needing no power so they function in the absence of emergency generating power) and some are active. In addition, the containment structure (see Figure 7.3) is designed to prevent any escape of radioactive substances even if the primary containment were to fail or leak. As described in Section 7.1.3, extensive multiphase flow analyses and simulated experiments (see, e.g., Hochreiter 1985) have been carried out to evaluate the effectiveness of these cooling systems following a postulated LOCA.

### 7.4.1 PWR Safety Systems

In a PWR, the ECCS (see Figure 7.2) consists of a number of water injection and spray systems, each with multiple injection points. There is a passive accumulator injection system consisting of two or more large tanks of water connected via a check valve to the primary coolant cold leg and maintained under a nitrogen pressure of 15–50 atm so that they inject water when the pressure in the primary coolant loop drops below a critical level. There are also several active water injection systems, typically a high-pressure coolant injection system (HPCI) designed to operate when the primary coolant loop pressure is high and therefore to operate for small breaks. There is also a low-pressure coolant injection system (LPCI) designed to operate for large breaks or when the primary coolant loop pressure is low. These injection systems are intended to flood the reactor core from below.

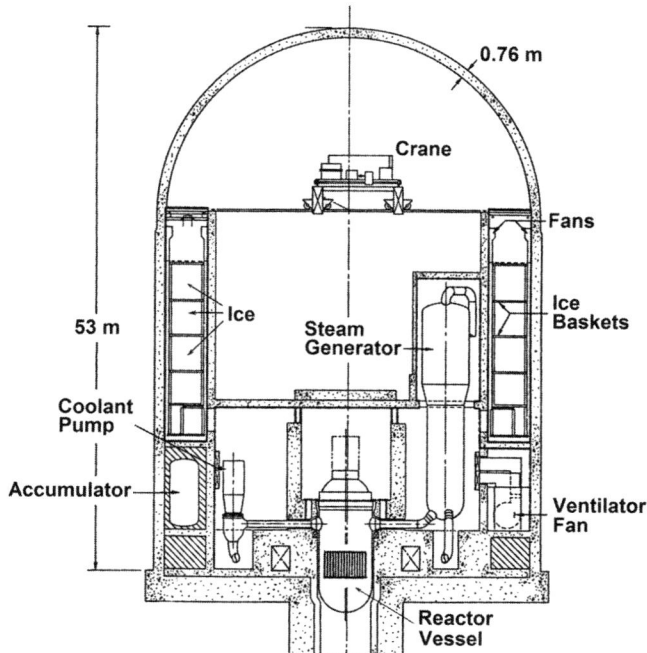

Figure 7.3. Typical PWR primary coolant loop and containment system. Adapted from USAEC (1973).

In a PWR, the secondary containment structure (see Figure 7.3) is designed to withstand the pressure that would be generated if all of the primary cooling water were released into that containment, a circumstance that is estimated to result in a maximum possible pressure of 5 atm. As shown, it is also equipped with cold water spray systems (see Figure 7.2) and, sometimes, ice to prevent the buildup of excessive heat and pressure within that containment in the event of cooling water and other substances escaping from the primary containment.

### 7.4.2 BWR Safety Systems

A typical BWR ECCS (Figure 7.4) has similar HPCI and LPCI systems as well as spray systems above the core and within the reactor vessel itself (see Figure 7.5). Usually one spray system is designed to operate while the pressure within the reactor vessel is high (the high-pressure core spray, HPCS) and another for lower reactor vessel pressures (the low-pressure core spray, LPCS). There is also a spray system outside the reactor vessel and inside the secondary containment structure whose purpose is to cool the primary containment vessel and its contents from the outside.

In a BWR, the potential consequences of the release of all of the primary cooling water are handled differently than in a PWR. As described in Section 6.6.4, the steam would be forced down into a *pressure suppression pool* or *wetwell*, where it would condense and thus prevent a buildup of pressure in the primary containment. The first Mark I configuration of this suppression pool was a toroidal shape, as shown

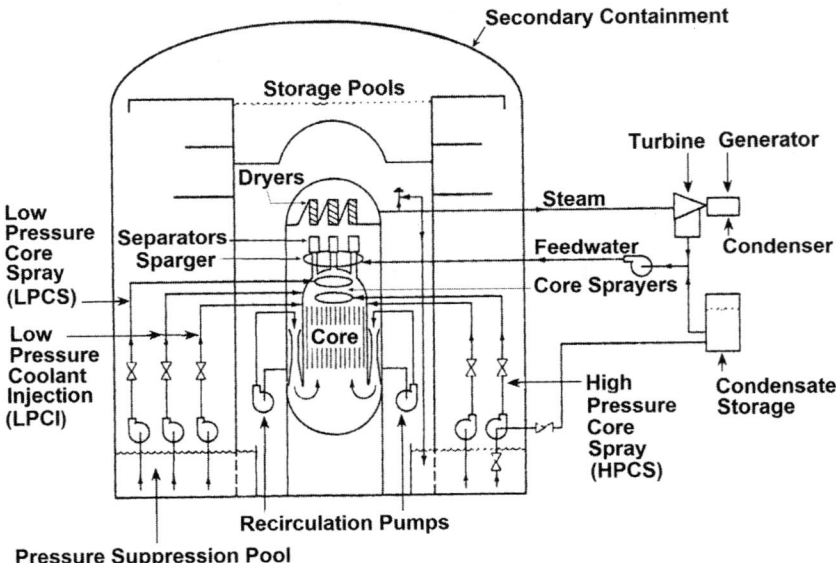

Figure 7.4. Schematic of the ECCS system in a GE Mark III BWR. Adapted from Dix and Andersen (1977) and Lahey (1977).

in Figure 7.6; General Electric introduced later Mark II and Mark III versions that are sketched in Figure 7.7. Concerns about the oscillatory condensation phenomena that might occur if these suppression pools were to be brought into action (see Section 6.6.4) raised issues of the structural loads that might result and the ability of the suppression pool structure to withstand those loads. Several very large scale experiments were carried out to answer those questions.

## 7.5 Major Accidents

There have been three major accidents at nuclear power stations. Each of these has not only had a major political effect on the future of nuclear power but has also driven home some important lessons that have greatly improved the safety of nuclear power plants. The political implications are beyond the scope of this text, though it is clear that they will cause any future developments in the industry to be very conservative. For example, it is hard to visualize that any reactor with positive or near-positive void coefficient of reactivity (see Section 7.1.2) would be politically acceptable in the foreseeable future, and this may eliminate many FBR designs.

The engineering lessons learned are, however, within the scope of this text and demand a description of all three of the major accidents. Of course, there have also been a number of lesser accidents, and mention of these are made where appropriate.

### 7.5.1 Three Mile Island

In March 1979 the operational PWR at Three Mile Island experienced a LOCA (see Sections 7.6.2 and 7.6.3) when a pressure relief valve in the pressurizer (Figure 4.5)

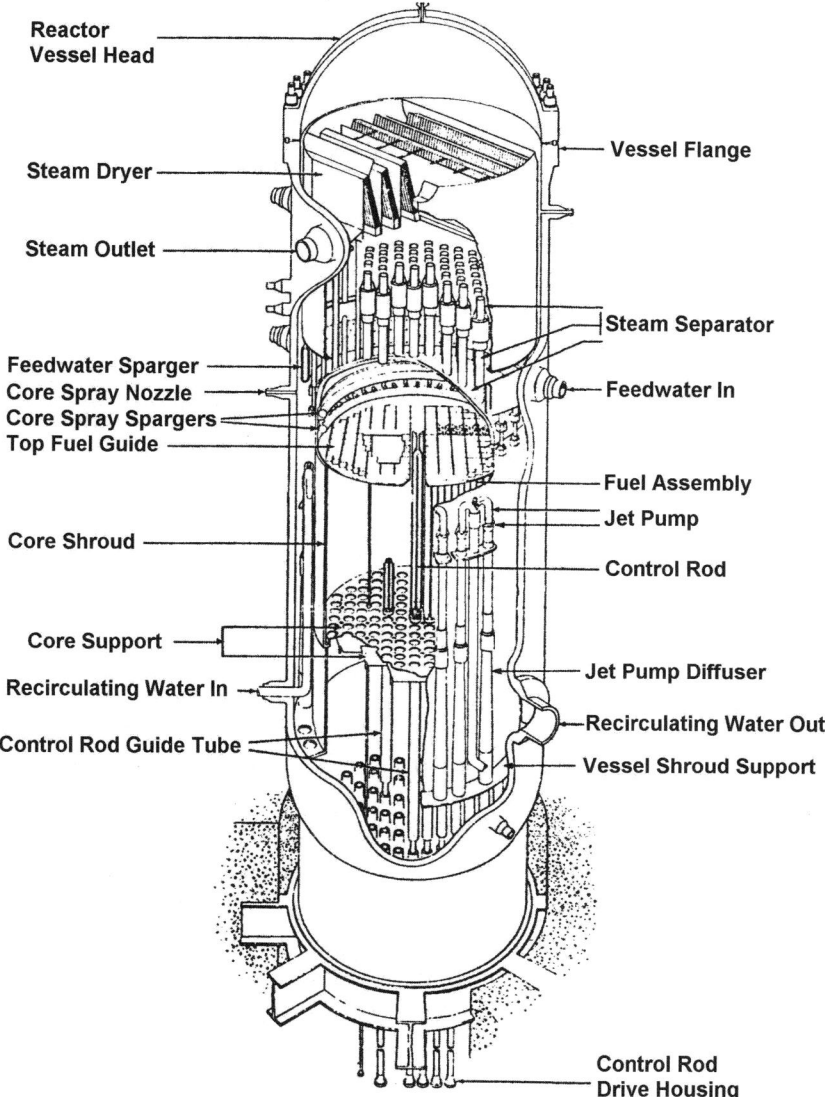

Figure 7.5. Typical BWR reactor vessel. Adapted from USAEC (1973).

stuck open without the operators realizing what had happened (Cameron 1982). The primary coolant drained out of the core that then overheated. The operators injected emergency cooling water with little effect partly because, unknown to them, water continued to drain out of the jammed pressure relief valve. Meanwhile, unexpectedly, a large bubble of steam and gas formed at the top of the core and prevented water from rising into it and cooling it. Half of the reactor melted and, in the process, the operators were forced to release a little radioactive steam to the atmosphere to prevent excessive pressure buildup in the containment building. Parenthetically, there was some buildup of hydrogen due to the high-temperature interaction of steam with the zircaloy cladding, and this may have exploded in the upper core. Eventually,

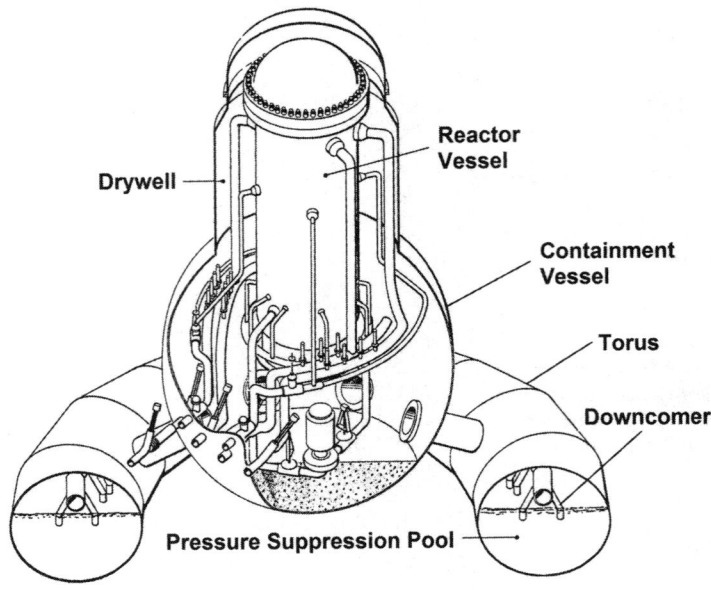

Figure 7.6. Schematic of the BWR (Mark I) primary containment and pressure suppression systems. Adapted from USAEC (1973).

sufficient water was forced into the core to cool it and bring the situation under control. The reactor's other protection systems functioned as they should, and the concrete containment building prevented any further release of radioactive material.

For some months after the accident, it was assumed that there had been no core meltdown because there was no indication of serious radioactive release within the secondary containment structure. However, as depicted in Figure 7.8 (IAEA 2015; see also Osif et al. 2004), it transpired that almost half the core had melted. Despite this, the reactor vessel remained almost completely intact and there was no major escape of radioactive material into the secondary containment structure. This helped allay the worst fears of the consequences of core meltdown in other LWR plants.

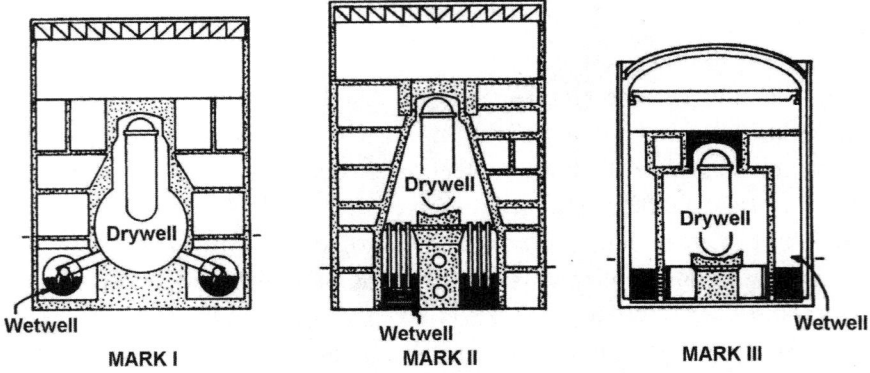

Figure 7.7. Mark I, Mark II, and Mark III BWR pressure suppression systems. Adapted from Lahey (1977).

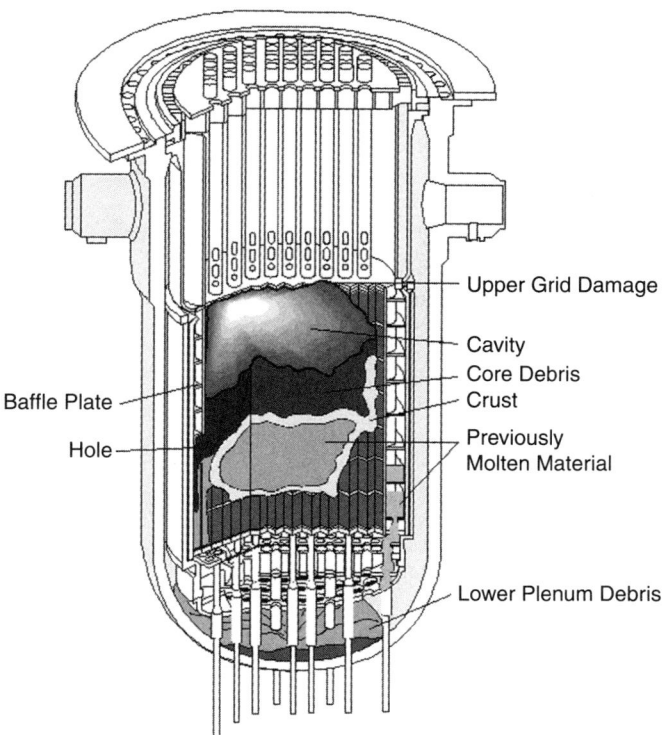

Figure 7.8. State of the Three Mile Island reactor after the accident (IAEA 2015).

The principal conclusion in the aftermath was that improved instrumentation was needed to ensure the operators had reliable information on the state of the reactor systems. If they had known the relief valve was open, the damage to the reactor would have been much less. In addition, it was concluded that operator error also contributed to the accident, and therefore improved training was also needed.

### 7.5.2 Chernobyl

The worst nuclear reactor accident occurred in the Ukraine in April 1986 when an old Russian RBMK-1000 boiling water reactor (see Figure 4.14) suffered an intense explosion and fire in the nuclear core. The accident and its aftermath have been extensively documented (see, e.g., Knief 1992; Marples 1986; Mould 2000) and exhaustively analyzed.

The accident occurred during a test carried out to determine the feasibility of using energy from the turbine coastdown during a reactor scram as a source of emergency electrical power. The idea was to eliminate the need for costly emergency power systems that would have required either continuously operating diesel generators (the generators could not be started quickly enough to meet the need) or an independent auxiliary cooling system. The plan involved initiating the test at a reduced power level with control rods partially inserted and the bypassing of some safety systems. It was assumed that the start of the reactor trip and the turbine

Figure 7.9. Photograph of the Chernobyl accident site taken shortly after the accident (USNRC 2006).

shutdown would coincide. However, a substantial delay in the reactor shutdown due to electricity needs caused a buildup of xenon poison, and this was counteracted by substantial control rod withdrawal. Shortly thereafter, all the primary cooling pumps were activated to ensure adequate cooling after the test. This, in turn, increased the heat transfer, essentially eliminated boiling of the coolant, and removed the reactivity margin that might have resulted from the boiling. The combination of low power and high flow led to instability that the operators had trouble controlling. A short time later, the planned test was initiated, the power began to rise, coolant voiding increased, and, recognizing the potential consequences, the operators began insertion of all control rods. However, the displacement of the coolant this produced led to increased reactivity and a huge surge in the power level. This was sufficient to cause fuel disintegration and a breach in the cladding that caused a huge steam explosion (see Section 7.6.4) that lifted the top off the reactor, blew off the building roof, and sent a plume of radioactive gases and particulates high into the atmosphere. The intense fire in and exposure of the nuclear core not only resulted in destruction of the reactor and the death of 56 people but also caused radiation sickness in another 200–300 workers and firemen. It also contaminated a large area in Ukraine and the neighboring country of Belarus (see Figure 7.10). It is estimated that 130,000 people in the vicinity of the reactor received radiation above international limits. The photograph in Figure 7.9 demonstrates how extensive the damage to the reactor building was.

This reactor not only did not have a secondary containment structure that might have prevented much death, injury, and contamination but it was also of the type

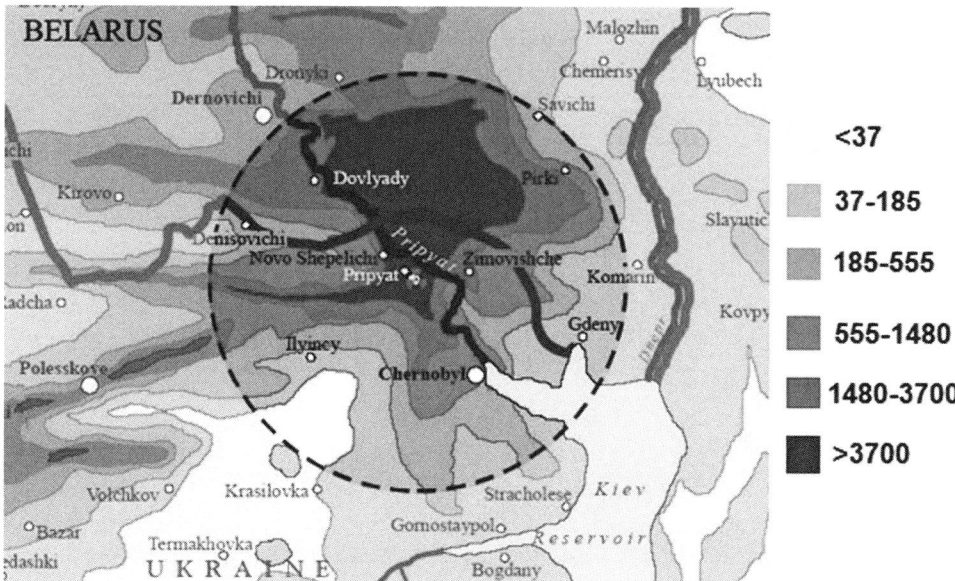

Figure 7.10. Map of the radioactive deposits (radioactivity of $^{137}$Cs in the soil in kBq/m$^2$) on April 27, 1986, and the 30 km exclusion zone around the Chernobyl reactor (UNSCEAR 2000).

that could have a positive void coefficient (see Section 7.1.2) that may or may not have been a factor during the leadup to the fuel disintegration. Eventually, with great difficulty and with considerable risk to human life, the remains of the reactor were covered in concrete. Plans to enclose the whole mess with an additional 107-m-tall, semicylindrical containment building that will be slid over the top of the damaged reactor building are currently under way (see Figure 7.11).

The Chernobyl disaster demonstrated the serious safety deficiencies in these old Russian nuclear power plants. As in the case of the Three Mile Island accident, the

Figure 7.11. Cylindrical entombment structure (left) being prepared for installation over the damaged Chernobyl reactor (center). Photograph reproduced with the permission of the owner, Ingmar Runge.

Figure 7.12. Tsunami striking the Fukushima Daiichi nuclear power plant on March 11, 2011. Reproduced with permission of Tokyo Electric Power Company (TEPCO 2011).

sequence of events that led up to the power surge were not adequately anticipated, and the bypassing of some of the safety systems may have been a contributing factor. Most notably, the lack of several layers of reactor confinement, particularly a carefully designed secondary confinement structure, led to much more severe consequences than might otherwise have been the case. These old reactors have now been removed from service or radically altered, and similar hypothetical mishaps have been carefully analyzed to ensure that there could be no repeat of the Chernobyl disaster.

### 7.5.3 Fukushima

On March 11, 2011, three operating Mark 1 BWRs at a power station in Fukushima, Japan (three out of the six at the site – the other three were not operating), shut down automatically and successfully when they experienced a huge magnitude 9.0 earthquake. One hour later, cooling, driven by the backup generators, was proceeding normally when the generators were swamped by a large tsunami (see Figure 7.12), causing the generators to stop and the ECCS systems to fail. The cores heated up uncontrollably and partially melted before the situation was brought under control, though not before several hydrogen explosions occurred. Despite this series of failures, the secondary containment was largely successful. There were no deaths, though some workers received nonlethal radiation doses. Figure 7.13 shows the evacuation area and radiation levels after 7 months (WNA 2014) near the damaged Fukushima Daiichi nuclear power plant.

This accident did confirm the viability of the secondary containment system but showed that more attention needed to paid to the siting of nuclear power plants

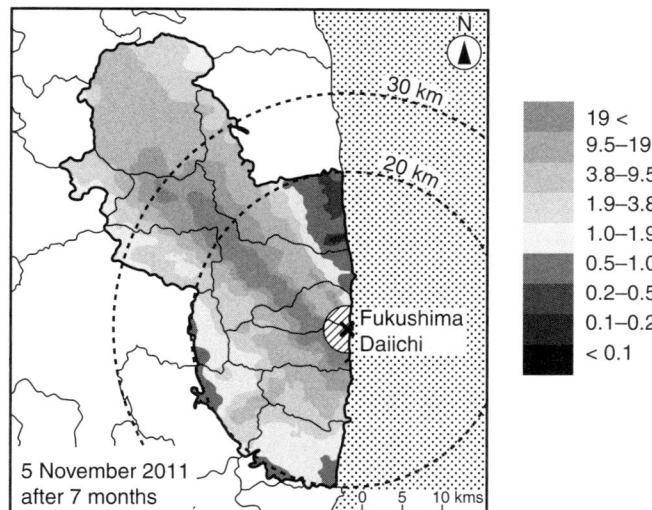

Figure 7.13. Fukushima Daiichi nuclear accident consequences: evacuation area (black outline) and radiation levels after 7 months measured 1 m above ground including background (in $\mu$Sv/hr). Adapted from WNA (2014).

(see Okrent 1981) and to the emergency provision of power for the safety systems at nuclear power plants. Many plants are situated near large bodies of water so as to provide cooling water; therefore they may be susceptible to floods, tides, and tsunamis. Upgrading of the protection from such hazards has been occurring around the world, most notably in France. Moreover, in the aftermath of Fukushima, renewed attention has been given to (1) safety systems that are passive in the sense that they do not need emergency power and (2) improvements to the protection of the power supply for emergency systems.

### 7.5.4 Other Accidents

There have, of course, been other lesser accidents during about 15,000 cumulative reactor-years of commercial nuclear power plant operation throughout the world. There has also been extensive experience in other reactors, mostly military and experimental. One of the worst accidents in other reactors was the Windscale fire of October 10, 1957, in one of two graphite-moderated air-cooled processing reactors at the Windscale facility on the northwest coast of England (Windscale Accident 2015). The two reactors had been hurriedly built as part of the British atomic bomb project. The fire occurred when one of the fuel channels overheated and caught fire; it burned for three days and caused a release of radioactive material that spread across the United Kingdom and Europe. In particular, this led to much concern regarding the spread of the radioactive iodine isotope [131]I and the contamination of milk in particular. Though no evacuation took place, dairy produce was destroyed for about a month.

There has also been extensive experience in other electricity-generating reactors, mostly military; in particular, the U.S. Navy, which has operated nuclear power plants since 1955, has an excellent safety record. Among the nonmilitary plants, aside from the three major accidents, there have been approximately 10 core meltdowns, mostly in noncommercial reactors, and none of these generated any hazard outside the plant. One of the reasons for the fine record of the U.S. Navy is that there was broad standardization in the design, construction, and management of its nuclear power plants (though two USN nuclear submarines have been lost for other reasons, and there have been reactor accidents including LOCAs in Soviet and Russian nuclear submarines; Johnston 2007). This allowed for safety experience to be broadly applied with subsequent widespread benefit. It is now recognized that a corresponding lack of standardization in commercial power plants significantly impaired their safety margins. In the aftermath, both national and international agencies charged with nuclear plant oversight are actively involved in pressing for standardization not only in the construction of new plants but also in the upgrading of older plants. In a broader context, global cooperation on safety issues has increased greatly in the aftermath of Chernobyl and Fukushima (see, e.g., OCED).

## 7.6 Hypothetical Accident Analyses

Safety concerns with nuclear plants, particularly the fear of the release of radioactive materials, have led to very careful analyses of all the conceived deviations from normal operation of the reactor and of all the conceived accidents that might have serious consequences. Of course, about 15,000 reactor-years of accumulated experience with nuclear power generation around the world have contributed substantial validity to these conceivable accidents and their likelihood of occurrence. One of the lessons from this experience is that the combination of minor events can sometimes lead to major problems. This makes accident prediction even more complex because it requires investigation of many more accidental permutations.

Conceivable events in a nuclear generating plant are classified as (1) normal operating transients that require no special action, (2) faults that may require reactor shutdown but that allow fairly rapid return to normal operation, (3) faults that result in unplanned shutdown that will result in extended shutdown, and (4) limiting faults that may result in the release of radioactive material.

### 7.6.1 Hypothetical Accident Analyses for LWRs

Addressing first the last category of faults (limiting faults) in light water reactors (see USAEC 1973; USNRC 1975), this category includes but is not confined to the following postulated accidents:

1. major rupture of the primary coolant loop pipes (PWR and BWR) leading to a LOCA
2. major rupture of the secondary coolant loop pipes (PWR)

3. steam generator rupture (PWR)
4. locked rotor on a coolant pump
5. fuel handling accident
6. failure of control rod mechanism housing
7. tornadoes, flooding, earthquakes, and so on

The focus here is on the first item, the LOCA, because it has been judged the postulated accident most likely to lead to the release of radioactive material.

### 7.6.2 Loss-of-Coolant Accident: LWRs

The worst scenario leading to a LOCA envisages an instantaneous double-ended or guillotine break in the primary coolant piping in the cold leg between the primary containment vessel and the primary coolant pump. This would result in the rapid expulsion of reactor coolant into the primary containment, loss of coolant in the reactor core, and rapid increase of the temperature of the core. This, in turn, might lead to a rapid increase in the pressure and temperature in the secondary containment; consequently, the secondary containment must be designed to withstand these temperatures and pressures as well as potential complications that might follow (see later). Moreover, even though the loss of coolant in the core would result in shutdown of the chain reaction (see Section 7.1.2), the decay heat could result in core meltdown unless the emergency core cooling systems were effective. Core meltdown might result in radioactive materials being released into the secondary containment, and hence that secondary barrier needs to be designed to contain those radioactive materials.

The progress of a hypothetical LOCA and the steps taken to bring the accident under control can be divided into three phases, namely, the blowdown phase, the refill phase, and the reflood phase. During the first or blowdown phase, the coolant is visualized as flashing to steam with two-phase flow proceeding through the primary cooling system and out through the guillotine break. Such multiphase flows are not easy to simulate with confidence, and much effort has gone into developing computer codes for this purpose (see Section 7.1.3) and into experimental validation of the results of those codes. These validation experiments needed to be conducted at large scale due to the uncertainty on how these multiphase flows scale (see, e.g., Holowach et al. 2003; Grandjean 2007). To evaluate the behavior of the multiphase flow in a PWR LOCA, a large-scale facility called the *Loss of Fluid Test Facility* (LOFT) was constructed at the Idaho National Laboratory. Advantage was also taken of a decommissioned reactor structure at Marviken, Sweden, to conduct additional blowdown tests mimicking a LOCA. For BWRs, General Electric conducted special full-scale blowdown tests at Norco in California. Key outcomes from these experiments were estimates of (1) the rate of steam and enthalpy ejection from the primary containment, a process that probably involved critical or choked flow through the effective orifice created by the break; (2) the forces placed on the system by this flow to evaluate the possibility of further structural damage; and (3) the amount of heat removed from the core

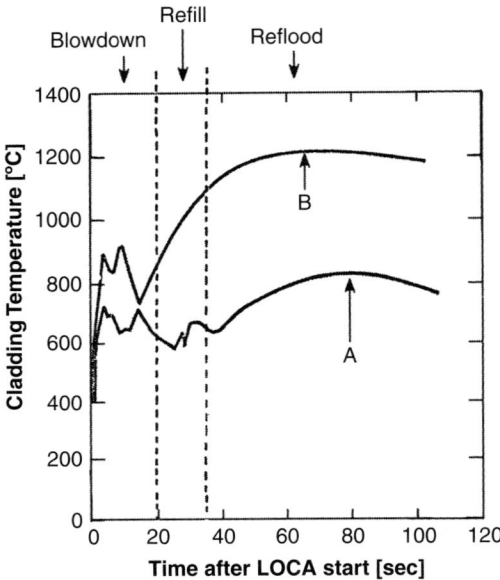

Figure 7.14. Estimated maximum temperature in the cladding during a postulated LOCA in a PWR as a function of time: (A) using realistic assumptions and (B) using conservative assumptions. Adapted from Hsu (1978).

by this flow that, in turn, defines the role of the subsequent refill and reflood phases (some analyses assume, conservatively, that no heat is removed).

About 10–20 s after the start of the blowdown, the emergency core cooling system (ECCS) (described in Section 7.4) begins operation, and this marks the beginning of the second or refill phase. Accurate prediction of the complex two-phase flows generated by the injection and spray systems is essential to ensure that the accident can be brought under control. This relies on a combination of well-tested computational tools backed up by both small- and large-scale experiments. Using these tools, predictions can be made of the development of the LOCA and its amelioration. An example of the information obtained is presented in Figure 7.14, which shows how the maximum temperature in the cladding might change during the three phases of the accident using either conservative assumptions or best estimates.

By definition, the refill stage ends when the liquid coolant level in the lower plenum rises to the bottom of the core; the last or reflood stage begins at this time. Reflood involves the quenching of the hot core as the liquid coolant rises within it (see, e.g., Hochreiter and Riedle 1977). The liquid coolant may be coming from the spray and injection system above the core or from the injection below the core. In the former case, quenching may be delayed as the water is entrained by the updraft of steam originating either in the core or in the lower plenum as a result of continuing flashing of the coolant. Such a *countercurrent flooding condition* (CCFL) (see Brennen 2005) may delay quenching either throughout the core or only in the hotter central region of the core. Indeed, a strong steam circulating flow is likely in which a steam–water droplet flow rises in a central column of the core and descends outside this central region. Other important differences can be manifest during reflood. For example, the *fast reflood* is defined as occurring when the liquid velocity exceeds the quench front velocity at the surface of the fuel rods (typically about 0.04 m/s), whereas a *slow reflood* involves coolant velocities less than the quench front velocity. Consequently, the two-phase flow conditions during reflood are unsteady, complex, and three-dimensional and require substantial computational and experimental efforts to anticipate their progress.

### 7.6.3 Loss-of-Coolant Accident: LMFBRs

Studies of postulated LOCAs in LMFBRs necessarily begin with the two basic differences between LMFBRs and LWRs. The first and most obvious is that the coolant in the LMFBR (attention here is confined to sodium coolant) is contained at low pressure and at a temperature well below its boiling temperature. Consequently, a primary coolant loop depressurization does not lead to the kind of rapid vaporization that occurs during the initial phase of a LOCA in a LWR. However, the second major difference is that in most LMFBR designs, overheating of the coolant in the core that leads to boiling and increased void fraction then produces an increase in the reactivity and therefore to an increase in the heat generated. Accident analyses and safety systems necessarily take into account these major differences in the reactor designs.

Specifically, boiling and loss of sodium in the core of an LMFBR would cause changes in the reactivity, as follows. The sodium would no longer slow down the neutrons, and hence there would be proportionately more fast neutrons. The neutron absorption by the sodium would be absent, but this is a lesser effect than the increase in the number of fast neutrons. The net effect is an increase in the reactivity of the reactor, giving it a *positive void coefficient* (see Section 7.1.2), though, to some extent, this potential increase is reduced by the increase in the flux of neutrons out of the reactor at its *edges*. In most designs, this is not sufficient to overcome the positive void coefficient of the bulk of the reactor, and the resulting reactivity increase would therefore result in an increase in the core heat production. This is in contrast to the LWR response and means that a LOCA in an LMFBR could have more serious consequences and could more readily result in a core meltdown. This is the reason for a focus on the *hypothetical core disassembly accident* discussed later. It is, however, valuable to point out that there have been efforts to redesign an LMFBR core to achieve a negative as opposed to a positive void coefficient. One way this could be done would be to change the geometry of the core and the blanket so that the negative effect of an increased leakage of neutrons as a result of the voidage more than negates the positive void effect in the bulk of the core (Wilson 1977).

The most likely scenario for a LOCA in an LMFBR is considered to be a blockage in one of the core coolant channels that leads to overheating in that channel, to boiling, and to increased void fraction in the core coolant. With a positive void coefficient, this might lead to escalating temperatures and to possible melting of the cladding of the fuel rods. While this series of events could be avoided by prompt reactor shutdown, nevertheless, the consequences of such a cladding melt have been exhaustively analyzed to understand the events that might follow. The conceivable scenarios are termed *hypothetical core disassembly accidents* (HCDA) and, within that context, it is possible that a *vapor explosion* or a *fuel–coolant interaction* (FCI) event might occur. These phenomena are discussed in the sections that follow.

### 7.6.4 Vapor Explosions

One of the accident scenarios that is of concern and that has been studied in the context of both LMFBRs and LWRs is the possibility of a *vapor explosion*. To assess the potential for and consequences of a vapor explosion (or of a fuel–coolant interaction, as described in the section that follows), note must first be made of the basic classes of vaporization identified in Section 6.4.1. A vapor explosion is defined as the explosive growth of a vapor bubble(s) within a liquid due to the presence of a large, nearby heat source. As described in Section 6.4.1, explosive growth of this kind only occurs under a set of particular conditions when the growth is not limited by thermal or heat transfer effects but only by the inertia of the surrounding liquid that is accelerated outward during the bubble(s) growth. Vapor explosions can occur in a number of other technological circumstances. Cavitation at normal pressures is an example of a vapor explosion caused by depressurization of a liquid (Brennen 1995). Vapor explosions also occur when one highly volatile liquid mixes with another at a higher initial temperature. One example of this occurs when liquid natural gas (or methane) is spilled into water at normal temperatures (Burgess et al. 1972) (this is a particular issue in LNG transportation accidents).

In other circumstances, the thermal boundary layer at the interface of the bubble(s) inhibits the supply to the interface of the necessary latent heat of vaporization. This is what happens when water is boiled on the stove at normal pressures, and this effect radically slows the rate of vaporization and the rate of bubble growth, as described in Section 6.4.2, in effect eliminating the *explosion*. Such thermally inhibited growth is manifest in many technological contexts, for example, in the growth of bubbles in the liquid hydrogen pumps of liquid-propelled rocket engines (Brennen 1994). Thermally inhibited growth tends to occur when the liquid–vapor is at higher saturation pressures and temperatures, whereas nonthermally inhibited growth tends to occur closer to the triple point of the liquid–vapor.

As described in Section 6.4.3, other factors that can affect whether explosive growth or thermally inhibited growth occurs are the conditions at the interface. If the thermal boundary layer is disrupted by instability or by substantial turbulence in the flow, then the rate of vaporization will substantially increase, and explosive growth will occur or be reestablished. Indeed, in a cloud of bubbles, the growth itself can cause sufficient disruption to eliminate the thermal inhibition. The vapor explosion would then be self-perpetuating.

However, at the kinds of normal operating temperatures for the water coolant in a LWR or the sodium coolant in an LMFBR, all bubble growth (in the absence of other effects, as described in the following section) would be strongly thermally inhibited (Brennen 1995) and highly unlikely to cause a self-perpetuating vapor explosion. To the author's knowledge, no such event has been identified in any nuclear reactor for power generation (see Fauske and Koyama 2002).

### 7.6.5 Fuel–Coolant Interaction

A fuel–coolant interaction (FCI) event is a modified vapor explosion in which a second material (a "hot" liquid or solid) is brought into close proximity to the vaporizing liquid interface and provides the supply of latent heat of vaporization that generates vapor bubble growth. It belongs to a class of vaporization phenomena caused by the mixing of a very hot liquid or solid with a volatile liquid that then experiences vaporization as a result of the heat transfer from the injected material. Of course, the result may be either relatively benign thermally inhibited vapor bubble growth or it may be explosive, nonthermally inhibited growth. Both have been observed in a wide range of different technological and natural contexts, the latter often being described as an *energetic* fuel–coolant explosion. Examples of such energetic explosions have been observed as a result of the injection of molten lava into water (Colgate and Sigurgeirsson 1973) or of molten metal into water (Lewis 1977). The key to energetic fuel–coolant explosions is the very rapid transfer of heat that requires substantial surface area of the injected liquid (or solid): fragmentation of the "hot" liquid (or solid) can provide this necessary surface area. The studies by Witte et al. (1973) and their review of prior research showed that such energetic explosions always appear to be associated with fragmentation of the injected "hot" material. Research suggests that an energetic fuel–coolant interaction consists of three phases: (1) an initial mixing phase in which the fuel and coolant are separated by a vapor film, (2) breakdown of the vapor film leading to greater heat transfer and vaporization rates, and (3) an explosive or energetic phase in which the fluid motions promote even greater heat transfer and vaporization. In this last phase, the explosive behavior appears to propagate through the fuel–coolant mixture like a shock wave.

Examples of reviews of the wide range of experiments on fuel–coolant interactions can be found in Witte et al. (1970) and Board and Caldarola (1977), among others. However, none of the experiments and analyses on sodium and uranium dioxide showed any significant energetic interaction, and most of the experts agree that energetic fuel–coolant interactions will not occur in liquid-sodium LMFBRs (Fauske 1977; Board and Caldarola 1977; Dickerman et al. 1976).

## 7.7 Hypothetical Accident Analyses for FBRs

Even though the possibility of an energetic fuel–coolant interaction can be essentially (though not completely) eliminated in the analyses of hypothetical accident analyses in a liquid sodium–cooled LMFBR, there still remain the questions of how the reactor core meltdown would proceed, of whether the containment would be breached, of whether radioactive materials could be released into the surroundings, and how the heat generated in the disassembled core would be dissipated (Wilson 1977). Studies and experiments on the core meltdown show that the resulting sodium–uranium mix in the reactor contains sufficient sodium to take away the decay heat by boiling for many hours while the decay heat declines. In this regard, the pool-type reactors are superior to the loop-type (see Section 4.8) because they contain more sodium. Moreover, large-scale experiments have shown that mixtures of boiling sodium and molten fuel and cladding can coexist for many hours without energetic interactions. Despite these reassuring studies, even the most remote possibilities must be explored to allay public fears regarding fast breeder reactors.

### 7.7.1 Hypothetical Core Disassembly Accident

Detailed analyses of hypothetical core disassembly accidents in LMFBRs have been conducted by Fauske (1976, 1977, 1981) and others. Much of this analysis begins with the hypothetical melting of the cladding that allows molten fuel to mix with the sodium coolant. As Wilson (1977) observes, the questions that necessarily follow are complex and difficult to answer. What is the potential for a fuel–coolant interaction involving the molten fuel, the coolant, and pieces of solid or liquid cladding? Does the cladding melting then progress to other parts of the core? Where does the fuel end up? Is there a physical argument that could be used to place a limit on the damage to the core? And, most importantly, does the reactivity increase or decrease during the various scenarios that follow? While many of these complex questions will need to be addressed, primary focus needs to be placed on the maximum possible accident, for public acceptance of LMFBRs will depend on the design of safety systems to contain such an accident. As with LWRs, computational analyses will need to be coupled with experimental programs to validate those predictions. For a comprehensive summary of these issues, the reader is referred to the review by Wilson (1977).

### REFERENCES

Aerojet Nuclear Company (1976). RELAP4/MOD5: A computer program for transient thermal-hydraulic analysis of nuclear reactors and related systems (3 volumes). *ANCR-NUREG 1335.*

Board, S. J., and Caldarola, L. (1977). Fuel coolant interactions in fast reactors. In *Symposium on the thermal and hydraulic aspects of nuclear reactor safety. Volume 2: Liquid metal fast breeder reactors*, pp. 195–222 (eds. O. C. Jones and S. G. Bankhoff). ASME.

Bodansky, D. (1996). *Nuclear energy: Principles, practices and prospects.* Springer-Verlag.

Brennen, C. E. (1994). *Hydrodynamics of pumps.* Concepts ETI/Oxford University Press.

Brennen, C. E. (1995). *Cavitation and bubble dynamics.* Oxford University Press.

Brennen, C. E. (2005). *Fundamentals of multiphase flow.* Cambridge University Press.

Burgess, D. S., Biordi, J., and Murphy, J. (1972). Hazards of spillage of LNG into water. U.S. Bureau of Mines, PMSRC Rep. 4177.

Cameron, I. R. (1982). *Nuclear fission reactors.* Plenum Press.

Colgate, S. A., and Sigurgeirsson, T. (1973). Dynamic mixing of water and lava. *Nature,* **244,** 552–55.

Collier, J. G., and Hewitt, G. F. (1987). *Introduction to nuclear power.* Hemisphere.

Dickerman, C. E., Barts, E. W., De Volpi, A., Holtz, R. E., Murphy, W. F., and Rothman, A. B. (1976). Recent results from TREAT tests on fuel, cladding and coolant motion. *Annals of Nuclear Energy,* **3,** 315–22.

Dix, G. E., and Andersen, J. G. M. (1977). Spray cooling heat transfer for a BWR fuel bundle. In *Symposium on the thermal and hydraulic aspects of nuclear reactor safety. Volume 1: Light water reactors* (eds. O. C. Jones and S. G. Bankhoff). ASME.

Fauske, H. K., (1976). The role of core disruptive accidents in design and licensing of LMFBR's. *Nuclear Safety*, **17**, 550–67.

Fauske, H. K. (1977). Liquid metal fast breeder reactor safety: An overview including implications of alternate fuel cycles. No. CONF-771120-22. Argonne National Lab., Ill.

Fauske, H. K. (1981). Core disruptive accidents. In *Nuclear reactor safety heat transfer*, pp. 481–94 (ed. O. C. Jones). Singapore: Hemisphere.

Fauske, H. K., and Koyama, K. (2002). Assessment of fuel coolant interactions (FCIs) in the FBR core disruptive accident (CDA). *Journal of Nuclear Science and Technology*, **39**, 608–14.

Frost, D., and Sturtevant, B. (1986). Effects of ambient pressure on the instability of a liquid boiling explosively at the superheat limit. *ASME Journal of Heat Transfer*, **108**, 418–24.

Grandjean, C. (2007). Coolability of blocked regions in a rod bundle after ballooning under LOCA conditions—Main findings from a review of past experimental programmes. *Nuclear Engineering and Design*, **237**, 1872–86.

Groeneveld, D. C., and Gardiner, S. R. M. (1977). Post-CHF heat transfer under forced convective conditions. In *Symposium on the thermal and hydraulic aspects of nuclear reactor safety. Volume 1: Light water reactors*, pp. 43–73 (eds. O. C. Jones and S. G. Bankhoff). ASME.

Hassan, Y. A. (1986). Analysis of FLECHT and FLECHT-SEASET reflood tests with RELAP5 MOD2. *Nuclear Technology*, **74**, 176–88.

Hewitt, G. F., and Collier, J. G. (1987). *Introduction to nuclear power*. Taylor and Francis.

Hochreiter, L. E. (1985). *FLECHT SEASET program final report*. NUREG/CR-4167.

Hochreiter, L. E., and Riedle, K. (1977). Reflood heat transfer and hydraulics in pressurized water reactors. In *Symposium on the thermal and hydraulic aspects of nuclear reactor safety. Volume 1: Light water reactors*, pp. 75–107 (eds. O. C. Jones and S. G. Bankhoff). ASME.

Holowach, M. J., Hochreiter, L. E., Cheung, F. B., Aumiller, D. L., and Houser, R. J. (2003). Scaling of quench front and entrainment-related phenomena. *Nuclear Engineering and Design*, **223**, 197–209.

Hsu, Y.-Y. (1978). *Two-phase problems in PWRs*. In *Two phase flows in nuclear reactors*, vol. 1, Von Karman Institute for Fluid Dynamics Vol. 1, 19–34.

Hsu, Y.-Y., and Graham, R. W. (1976). *Transport processes in boiling and two-phase systems*. Hemisphere/McGraw-Hill.

Hsu, Y.-Y., and Sullivan, H. (1977). Thermal hydraulic aspects of PWR safety research. In *Symposium on the thermal and hydraulic aspects of nuclear reactor safety. Volume 1: Light water reactors*, pp. 1–15 (eds. O. C. Jones and S. G. Bankhoff). ASME.

IEAA (2015). *Safety case study: The Three Mile Island accident*. http://www.iaea.org/ns/tutorials/regcontrol/assess/assess3233.htm.

Jackson, J. F., Liles, D. R., Ransom, D. H., and Ybarrondo, L. J. (1981). LWR system safety analysis. *Nuclear Reactor Safety Heat Transfer*, **12**, 415.

Johnston, R. (2007). *Deadliest radiation accidents and other events causing radiation casualties*. Database of Radiological Incidents and Related Events. See also http://en.m.wikipedia.org/wiki/Nuclear_submarine\#Accidents.

Jones, O. C. (ed.) (1981). *Nuclear reactor safety heat transfer*. Hemisphere.

Jones, O. C., and Bankhoff, S. G. (eds.) (1977a). *Symposium on the thermal and hydraulic aspects of nuclear reactor safety. Volume 1: Light water reactors*. ASME.

Jones, O. C., and Bankhoff, S. G. (eds.) (1977b). *Symposium on the thermal and hydraulic aspects of nuclear reactor safety. Volume 2: Liquid metal fast breeder reactors*. ASME.

Knief, R. A. (1992). *Nuclear engineering: Theory and practice of commercial nuclear power*. Hemisphere.

Lahey, R. T., Jr. (1977). The status of boiling water nuclear reactor safety analysis. In *Symposium on the thermal and hydraulic aspects of nuclear reactor safety. Volume 1: Light water reactors* (eds. O. C. Jones and S. G. Bankhoff). ASME.

Lewis, E. E. (1977). *Nuclear power reactor safety*. John Wiley.

Long, G. (1957). Explosions of molten aluminum in water: Cause and prevention. *Metal Progress*, **75**, 107–12.

Marples, D. R. (1986). *Chernobyl and nuclear power in the USSR*. St. Martin's Press.

Mould, R. F. (2000). *Chernobyl record: The definitive history of the Chernobyl catastrophe*. Institute of Physics.

OCED. (1996). *Implementing severe accident management in nuclear power plants*. Nuclear Energy Agency, OCED.

Okrent, D. (1981). *Nuclear reactor safety*. University of Wisconsin Press.

Osif, B. A., Baratta, A. J.; and Conkling, T. W. (2004). *TMI 25 years later: The Three Mile Island nuclear power plant accident and its impact*. Pennsylvania State University Press.

Schultz, M. A. (1955.) *Control of nuclear reactors and power plants*. McGraw-Hill.

TEPCO. (2011). Tokyo Electric Power Company. http://photo.tepco.co.jp/date/2011/201105-j/110519-01j.html.

Todres, N. E., and Kazimi, M. S. (1990). *Nuclear systems I. Thermal hydraulic fundamentals*. Hemisphere.

UNSCEAR (2000). Exposures and effects of the Chernobyl accident. *United Nations Scientific Committee on the Effects of Atomic Radiation* UNSCEAR 2000 Rep. **2**, Annex J.

USAEC (1957). Theoretical possibilities and consequences of major accidents in large nuclear power plants. *U.S. Atomic Energy Commission* Rep. WASH-740.

USAEC (1973). The safety of nuclear power reactors (light water cooled) and related facilities. *U.S. Atomic Energy Commission* Rep. WASH-1250.

USNRC (1975). Reactor safety study: An assessment of accident risks in U.S. commercial nuclear power plants. *U.S. Nuclear Regulatory Commission* Rep. WASH-1400.

USNRC (2006). Safety culture and changes to the ROP. Inspector counterpart training session, May–June 2006. http://pbadupws.nrc.gov/docs/ML0615/ML061590145.pdf.

Wagner, R. J., and Ransom, V. H. (1982). RELAP5 nuclear-plant analyzer capabilities. *Transactions of the American Nuclear Society*, **43**, 381–82.

Wilson, R. (1977). Physics of liquid metal fast breeder reactor safety. *Review of Modern Physics*, **49**, 893–924.

Windscale Accident (2015). http://en.wikipedia.org/wiki/Windscale_fire.

Witte, L. C., Cox, J. E., and Bouvier, J. E. (1970). The vapour explosion. *Journal of Metals*, **22**, 39–44.

Witte, L. C., Vyas, T. J., and Gelabert, A. A. (1973). Heat transfer and fragmentation during molten-metal/water interactions. *ASME Journal of Heat Transfer*, **95**, 521–27.

WNA. (2014). *Fukushima accident*. http://www.world-nuclear.org/info/Safety-and-Security/Safety-of-Plants/Fukushima-Accident/.

# Index

absorbing material, 67
absorption, 9, 11
  cross section, 11, 18
accidents, 134–42
actinides, 14
Aerojet-General, 65
AGR, 19, 69
airliner impact, 131
alpha radiation, 15
anisotropy, 36
annular flow
  instability, 98–99
Avagadro's number, 11

balance of plant, 5, 56
barns, 11
becquerel, 15
Bernoulli effect, 97
beta radiation, 15
blanket, 23, 39, 70, 71
blowdown, 143, 144
BN-600, 21, 71, 73
BN-800, 74
boiling, 86, 107–18
  crisis, 111
  vertical surfaces, 116–18
boric acid, 67
boron, 67
burnable poison, 58, 67
burnout, 89, 127
burnup, 8
BWR, 19, 60–62, 69, 86–90, 109, 127
  Fukushima, 140
  RBMK, 137

cadmium, 67
calandria, 68
CANDU reactor, 7, 19, 58, 67–69, 128
capture resonance, 18
carbide, 74
cavitation, 105–107, 146

CCFL, 144
centrifuge, 7
chain reaction, 8, 18, 19, 21, 68, 70, 143
Chernobyl, 1, 69, 130, 131, 137–40
CHF, 111, 127
*China Syndrome, The*, 130
chugging, 120–24
cladding, 7, 8, 16, 62, 80
Clausius–Clapeyron equation, 107
Clinch River, 74
concentration wave, 94, 119
condensation
  instability, 120–24
  oscillations, 121
contact resistance, 79, 80
containment, 16–17, 59, 132, 136
  primary, 70, 132, 143
  secondary, 61, 62, 130, 133, 136, 138, 140,
    143
continuous phase, 93
control blade, 63
control rod, 43–46, 62–64, 67
  channel, 63, 67
  insertion, 45, 83, 131, 137, 138
conversion, 7
coolant
  $CO_2$, 69
  fluoride salt, 75
  heat transfer, 82–83
  helium, 70, 75, 76
  lead, 76
  liquid metal, 70
  lithium, 71
  molten salt, 75
  primary loop, 59, 60, 71, 99, 143
  pump, 143
  secondary loop, 59, 60, 70, 142
  sodium, 71, 74, 76
  supercritical water, 76
cooling systems, 132, 137
  passive, 132

cooling water, 17, 135
core design
    LMFBR, 85–86
    LWR, 84–85
    pebble bed, 75
    prismatic block, 75
core disassembly, 145, 148
core meltdown, 128, 130, 136, 140, 143, 145, 147
critical heat flux, 86, 87, 89, 111, 114, 127, 128
critical size, 21
criticality, 21–22, 34, 67
cross section, 11, 30
curie, 15
current density, 26
cylindrical reactor, 38–39, 85

decay heat, 8, 15, 78, 132, 143, 147
delayed neutrons, 13, 53, 66, 67
    precursor, 13
density wave, 119
diffusion coefficient, 30, 33
diffusion equation
    one-speed, 32, 33
diffusion theory, 30–32
    one-speed, 32–34
    two-speed, 34–36
disperse flow, 93
    friction, 99–101
    horizontal, 99–100
    limits, 95–96
disperse phase, 93
dispersion, 96
driver, 23, 71

earthquakes, 131, 140
ebullition cycle, 112
ECCS, 132, 140, 143, 144
    BWR, 133
    PWR, 132
elastic scattering, 9
electricity generation, 2
electron volt, 9
embrittlement, 17
enrichment, 7
    tailings, 7
entrance length, 92
epithermal neutrons, 19
escape probability, 20
extrapolation length, 32

fast breeder reactor, 21
    liquid metal, 21
fast fission factor, 20
fast neutrons, 9, 18
fast reactor, 21
fault tree, 130
FBR, 21, 70, 134
    accident analyses, 147–48

FCI, 146
fertile
    isotope, 18
    material, 7, 21, 23
Fick's law, 29, 30
film boiling, 86, 115–16
fissile
    atoms, 5
    isotope, 18
    material, 6, 21
fission, 5, 9
    cross section, 11, 18
    fragments, 8, 9
    products, 8, 78
FLECHT program, 129
flooding, 113
flow
    pattern, 91
    regime, 91
flow regimes
    annular flow, 93
    churn flow, 93
    classification, 93–95
    disperse flow, 93
    map, 92–93
    slug flow, 93
FNR, 70
four-factor formula, 20
friction
    coefficient, 100
    interfacial, 104
    pipe flow, 99
fuel
    assembly, 62, 71
    bundle, 62, 67, 68
    heat transfer, 79–82
    pellets, 7, 62, 67, 79
    rod, 7, 46, 62, 79–82
    temperature, 79, 82, 128
    tubes, 67
    volume, 85
fuel–coolant
    interaction, 105, 146–48
Fukushima, 130, 140–41
fusion, 5

gadolinium, 67
gamma radiation, 15, 17, 78
gas-cooled reactor, 19
gaseous diffusion, 7
GCR, 19, 76
Generation II, 56
Generation IV, 56, 74–76
    fast reactors, 76
    thermal reactors, 75
geometric buckling, 22, 34
GFR, 76
graphite, 19
    moderator, 69, 70, 75, 76

Haberman–Morton number, 114
half-life, 13
HCDA, 145
heat conduction, 80
heat exchanger, 59
heat production, 79, 81
heat release, 78
  delayed, 78
  prompt, 78
heat transfer, 78–89
  coefficient, 82
heavy water, 19, 58, 67
  reactor, 19
Helmholtz's equation, 34
homogeneity, 37, 45
homogeneous flow
  friction, 100–101
HPCI, 132, 133
HPCS, 133
HTGR, 7, 19, 70, 76
hurricanes, 131
HWR, 19, 67–69
hydraulic diameter, 82, 102

inelastic scattering, 9
inhomogeneity, 37
interfacial roughness, 108

Kelvin–Helmholtz instability, 96–98,
    119
kinematic wave, 94, 119

latent heat, 86
lattice cell, 37, 45–51, 63
  control rod, 49
  fuel rod, 47
  square, 51
  theory, 37
Ledinegg instability, 119–20
LFR, 76
light water, 19, 58, 67
  reactor, 19
LMFBR, 21, 23, 70–74, 128, 147
  loop-type, 71, 147
  pool-type, 71, 76, 147
LOCA, 89, 128, 130, 132, 134, 142–46
  LMFBR, 145–46
LOFT, 129, 143
loss of coolant, 130, 131, 134, 142–46
LPCI, 132, 133
LPCS, 133
LWR, 19, 57–62, 76
  accident analyses, 142–45
  control, 66–67

macroscopic cross sections, 12
Manhattan project, 18
Martinelli correlations, 101
Martinelli parameter, 103

mass
  flux, 91
  fraction, 91
  quality, 86, 91, 105
material averaging, 27–28
material buckling, 22, 33
mean free path, 12
meltdown, 128
microlayer, 110
mixed oxide, 8
  fuel, 22, 23
  reactor, 8
mixture
  density, 91
  viscosity, 100
moderator, 18–20
  volume, 85
Monte Carlo methods, 54
MOX, 8, 22, 23
MSR, 75
multigroup
  diffusion model, 36, 37
  theory, 36–37
multiphase flow, 90–128
  flow patterns, 91–99
  flow regimes, 90
  fully separated flow, 94
  homogeneous flow, 93
  in nuclear reactors, 124, 127–29
  instability, 118–24
  intermittency, 94, 95
  notation, 90–91
  regime, 91–99
multiplication factor, 10, 20–22, 34, 52, 67, 128
multiscale models, 54

natural convection, 109, 116
neutron
  continuity equation, 29
  density, 25
  diffusion coefficient, 30
  diffusion length, 31, 33, 35
  energy, 10
  flux, 25, 26, 83
  flux distribution, 25, 41, 44–46, 48, 50, 52, 79, 83
  mean free path, 31
  transport theory, 27
  velocity, 25
nonleakage probability, 20
NSSS, 56
nuclear
  energy spectrum, 10–11
  explosion, 1, 131
  fission, 9–10
  fuel, 7
  fuel cycle, 5–9, 22–23
  fusion, 13
  waste, 1, 2
nucleate boiling, 109, 111–12

nucleation, 109
  sites, 111
NuScale, 65
Nusselt number, 82

one-speed model, 32

$P_1$ equations, 36
$P_N$ equations, 36
Phenix, 21, 71, 74
pipe friction, 99–105
plutonium, 8, 14, 70
point kinetics model, 53
pool boiling, 108–11
  crisis, 113–15
power outages, 131
Prandtl number, 82
pressure drop, 99–105
pressure suppression systems, 121
pressurizer, 59, 134
prompt neutrons, 10, 66
PWR, 19, 58–59

quality, 91
quality factor, 16

rad, 16
radiation, 8, 15–16, 138, 140
  attenuation, 17
radiative capture, 9
radioactive
  contamination, 59
  decay, 13–15
  decay constant, 14
  release, 16–17, 130–32, 135, 136, 138, 142, 143,
    147
  sodium, 71
radioactivity, 13–18, 59, 62, 130
Rayleigh scattering, 33
Rayleigh–Taylor instability, 97, 98, 108, 115
RBMK reactor, 69
reactivity, 10, 87, 128
  control, 58, 131
reactor
  boiling water, 60
  burner, 21
  critical, 21
  dynamics, 51
  epithermal, 75, 76
  fast breeder, 21
  fast neutron, 70
  gas-cooled, 69
  gas-cooled fast, 76
  heavy water, 67
  kinetics, 51
  lead-cooled, 76
  molten salt, 75
  natural, 6
  pebble bed, 75

period, 53
  personal, 65
  pressurized water, 58
  shielding, 17
  small modular, 65
  sodium-cooled, 76
  supercritical-water-cooled, 76
  thermal, 19
  vessel, 59
reduced order models, 54
reduced thermal models, 27
refill, 143, 144
reflector, 39–42
reflood, 143, 144
refueling, 67
RELAP code, 129
rem, 16
reprocessing, 2, 8
reproduction factor, 10
resonance peaks, 12
Reynolds number, 82, 96, 100
roentgen, 16

$S_N$ equations, 36
safety
  concerns, 1, 130–31, 142
  earthquakes, 131, 140
  systems, 131–34
  tsunamis, 131, 140, 141
safety systems, 131, 137, 140, 141
  LMFBR, 148
  passive, 141
  spray, 133, 144
scattering, 9, 11
  cross section, 11, 18
scram, 67, 137
scram control, 67
SCWR, 76
sedimentation, 96
separated flow, 93
  friction, 101–105
  limits, 96–98
SFR, 76
shielding, 17–18
shim control, 67
sievert, 16
six-factor formula, 20
slurry flow, 100
SMR, 65
spherical reactor, 37–38
standardization, 2
steam generator, 59, 143
steam supply system, 5
stratified flow, 98
subcritical, 21, 34
supercritical, 21, 34
supernova, 13
Superphenix, 21, 74
suppression pool, 61, 122, 133

tailings, 7
temperature distribution, 25, 80, 82–84
terrorist attacks, 131
thermal conductivity, 80, 81, 85
thermal cross section, 12
thermal diffusivity, 107
thermal fission factor, 20
thermal neutrons, 9
thermal shield, 17
thermal utilization factor, 20
thermo-hydraulics, 3
thorium, 7, 21, 22, 70
    fuel cycle, 7
Three Mile Island, 1, 74, 130, 134–37
transport theory, 28–30
transuranic elements, 8
tsunamis, 131, 140, 141
two-speed model, 32, 36

uranium, 5
    carbide, 70
    fuel, 62
    natural, 67
    ore, 6
    weapons grade, 9

vapor explosion, 105, 108, 138, 146
vaporization, 105
    classes, 105

heterogeneous, 108–18
homogeneous, 105–108
vertical flow
    friction, 101
VHTR, 75
void coefficient, 68, 128, 139, 145
void fraction, 87, 128
void reactivity, 134, 138, 145
volcanoes, 131
volume
    flux, 91
    fraction, 91
    quality, 91
volumetric flux, 91

waste, 10, 14
    disposal, 8
    storage pools, 8, 15
weakly absorbing medium, 31
Westinghouse SMR, 65
wetwell, 133

xenon poison, 138

yellowcake, 7

zircaloy, 16, 67, 135
    tubes, 62
zirconium, 16